AF714965

KISP
Prof. Kunow + Partner

Annette Kunow

Project Management

&

Business Coaching

Agile project management - target-oriented and efficient with active body language & comprehensive communication

1th english edition

NAME: Annette Kunow

ADRESSE: Baumhofstr. 39 d, 44799 Bochum

Web: www.kisp.de

E-Mail: info@kisp.de

Tel: 02349730006

Illustration: Annette Kunow

ISBN Nummer: 978-3-96695-006-0

PREFACE

Due to the very lean hierarchies in today's companies, the targeted management of smaller projects and the skillful leadership of people are the yardstick for project success. The interaction of these factors and the demands on the project manager are the subject of this textbook.

In order to make a project successful, the project not only has to be successfully led to its goal, but the "right" project has to be selected from the very beginning. For this purpose, project managers need marketing knowledge, which is explained by means of an example.

Another important factor is the leadership of the project staff. To achieve this, the project manager must have a good knowledge of human beings and leadership qualities and be able to use them in a targeted way to solve possible conflicts.

The interaction of these topics is described in detail. On the basis of current examples from practice, the project manager is given practical instructions to guide his project successfully. Practical exercises convey and deepen what you have learned.

Of course there is a discrepancy between theory and practice. It is clear that not all processes can be exactly implemented in practice, not least because project managers are often determined by others. Nevertheless, it is extremely important that every project manager is aware of the interrelationships in order to manage projects successfully.

Checklists and blank forms are used to provide suggestions for creating appropriate documents for your own project.

Finally, a short introduction to the Microsoft Project 2000? program is given, which is often used for small and medium-sized projects.

This book "Project Management & Coaching" is the result of practical work in industry, lectures at Bochum University of Applied Sciences and seminars in industry.

The target group is practitioners in industry and students of all disciplines who want to work as project managers in these areas.

In this eddition, I will go into the possible and recurring stumbling blocks and their causes in the project, in order to show the reader at an early stage what he or she has to pay attention to.

In addition, the aspects of agile project management will be further expanded. What opportunities do they offer? What are their advantages and disadvantages?

Additional bonus material that would have split the scope of this book, such as solutions to cases, checklists and much more, can be found on the Internet. The links are given at the places in the book.

Bochum, January 2019

Prof. Dr. -Ing. Annette Kunow

Here you can book a free Strategy Session or write me if you like this book and have suggestions or questions.

Click here for free bonus material on the book.

Also visit my blog „Selbstführung & Produktivität in German“. I'll help you get better results.

Table of Contents

1 INTRODUCTION

The first step is
halfway down the road.

1.1 Objectives of this Project Management & Business Coaching Book

Every person carries out projects every day. A student conducts a project as part of a student research project. A craftsman carries out a project to fulfill a sales order. A host conducts project-like social events. All of them can benefit from the same project management theory applied by engineers, team leaders and contractors. The principles are always the same.

This book will help you to carry out your projects successfully or even more successfully. It does not require any previous knowledge of management or technology.

The book will guide you from beginning to end of a project cycle and will address the main project phases and problems:

- the project definition,
- Project planning,
- the project organization,
- the project implementation,
- the project completion,
- the project management software.

Numerous questionnaires, checklists and exercises deepen the content and enable immediate use in practice. The listing of the relevant, cited literature allows a deepening of the book content.

But just as important as the structuring of a project is the leadership of people in the project team. The project manager must have a very good knowledge of human beings and leadership skills and be able to use them in a targeted way to solve possible conflicts. This is particularly demanding, because in most cases the project manager has no hierarchical competences vis-à-vis the team.

1.2 Your own learning objectives

Since the definition of the (project) goal is always at the beginning of a project, a goal definition is also necessary for the project "development of the content of a textbook".

The goal definition has two different components. On the other hand, each reader can set his own learning objective by, for example, using table 1.1 to develop a new project management capability.

Setting your own learning objective has two main advantages:

- while reading, your own learning goal allows you to keep an eye on the overall goal of the book, and
- if you actually reach the goal then you have a feeling of confirmation and satisfaction.

Chapter 3 shows how to define and formulate goals in general, including project goals, as precisely as possible. Finally, the creativity methods are discussed in detail in Chapter 9.

Table 1.1 shows some of the options that you can learn in project management & business coaching. All facets of the topic are addressed, the purely factual, but also the interpersonal ones. In practice, both aspects have approximately the same proportion.

These aspects are addressed where they occur in the course of the project, e. g."overcoming conflicts" in project implementation, etc.

First check the most important learning goals for you (maximum 5) and then check the skills you want to improve.

Table 1.1: Potential Project Management & Business Coaching Skills

In this project management & business coaching book I learn,		
1	Organize projects from start to finish.	☐
2	Structure plans that will withstand project pressure.	☐
5	To lead employees.	☐
6	Support employees with problems.	☐
7	To plan and optimise the use of existing resources.	☐
8	To organize a project within the given budget and time frame.	☐
9	Eliminate waste of time and money.	☐
10	Measure the progress and performance of a project.	☐
11	Appropriate use of information systems (project management software).	☐
12	Successful implementation of changes in the project process.	☐
13	Identify and resolve conflicts at an early stage.	☐
14	Define project goals comprehensively.	☐

2 What is Project Management

2.1 Development of Project Management

The concept of project management as a discipline was developed in the early 1960s to implement the American space programme that began at that time. Today, as already mentioned in chapter 1, smaller companies are also defined as projects.

Project management always focuses on one project.

The term project is defined as follows.

A project

- is limited in time,
- has a defined target,
- has a certain uniqueness,
- must usually be divided into sub-projects and individual activities (complexity),
- is generally associated with a risk,
- requires the cooperation of specialists from different fields for different time periods,
- requires project management and a project specific organization

The project is therefore a company with a clear objective and a definite beginning and end, which takes place uniquely (cyclically). It has a task (objective) which is carried out within the framework of certain personnel, budget, time and quality constraints. It is clearly different from other projects.

The skills and talents of the employees who work on this project, in whole or in part, but also other institutions (enterprises, institutions, etc.) and things such as tools, equipment, systems and procedures, and of course also financial means, i. e. the available money, are the resources of a project.

The term S. M. A. R. T. - project management originates from the english-speaking area and is used in project management as a criterion for the clear definition of goals within the scope of a target agreement /2.1/. S.M.A.R.T. is an acronym for "Specific Measurable Accepted Realistic Timely". It can be interpreted that way.

- S Specific: Objectives must be clearly defined. Not vague, but as precise as possible.
- M Measurable: Objectives must be measurable (measurability criteria).
- A Executable (accessible): Destinations must be accepted by the recipients. But they must also be appropriate, attractive or demanding /2.2/.
- R Realistic: Goals must be possible.
- T Terminable: A clear deadline for each goal is a prerequisite for when it must be reached.

A goal is S. M. A. R. T. if it meets these five conditions.

A project is generally characterised by a certain degree of complexity. However, this term is subjectively assessable. Today, smaller, less complex projects are also gaining in importance.

Project management refers to all management tasks, management organisation, techniques and means for the execution of a project. The term "project management" is unambiguously defined when leadership is understood as the control of the various individual activities in a project with regard to the project goals.

Project management thus becomes a comprehensive management concept that can be supported by individual management techniques and aids. These management techniques and tools are referred to in the following as business coaching and are described in detail.

The project management is distinguished from management in two fundamental respects

- Project management is limited to a clearly defined period of time, whereas the planning of a company organisation is based on the assumption that there is no time limit.
- Projects often require resources only on a part-time basis or in certain periods of time, while permanent organizations try to fully utilize their resources over the entire work year or depreciation period.

The second point often leads to conflicts within the company or between different departments: In many cases, project managers can only succeed in making the necessary resources available to the project, for example the most capable employees of a department, if they are able to negotiate with great skill.

In recent years, the "project-oriented company" has become increasingly important. It not only organizes the production of services for the customer or the market as a project, but also uses project management as a means of technical and organizational change within the company itself.

Due to the strong competitive pressure, many companies are restructuring their processes in such a way (e. g. reengineering or reorganization) that without project management the simultaneity of processes would no longer be possible (Fig. 2.1). Project management is therefore even more important.

2.2 Projects and Products

Definition

A product is the term for something developed or manufactured by a company, such as software, hardware or services.

These products are developed in projects. Projects are therefore used for product development.

The lifecycle of a product can also require several projects.

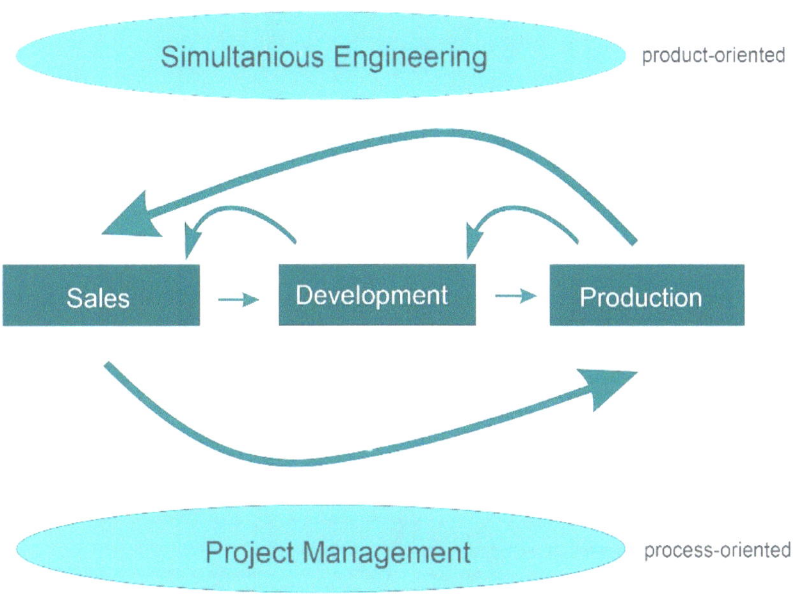

Figure 2.1 Simultaneous Engineering Processes

The difference between product and project management is therefore

- The product management is responsible for determining and determining the scope of development.
- Project management is responsible for the execution and delivery of the scope.

But product development must not be seen as a sequence of projects. Product management includes the following important tasks

- Contribute innovations
- Training personnel regularly

If innovations are not introduced, "technical" debts can accumulate, which are difficult to reduce later on.

If no new, valuable technical knowledge is built up, it can lead to organisational problems. Even qualified employees do not emerge in 9 months.

Product management must ensure that the product goes beyond the scope of the next release and can be kept marketable with innovations /2.4/.

2.3 The Life Cycle of a Project

Each project goes through a foreseeable, un-divided life cycle in four phases, which requires the project manager to have very different abilities in each phase:

- Phase 1: Concept and definition of the project
- Phase 2: Planning of the project
- Phase 3: Implementation of the plan
- Phase 4: Completion and evaluation of the project

These individual phases are explained in detail below.

In table 2.1 you can convince yourself of your skills as a project manager and work through a project you have carried out once again and then evaluate it.

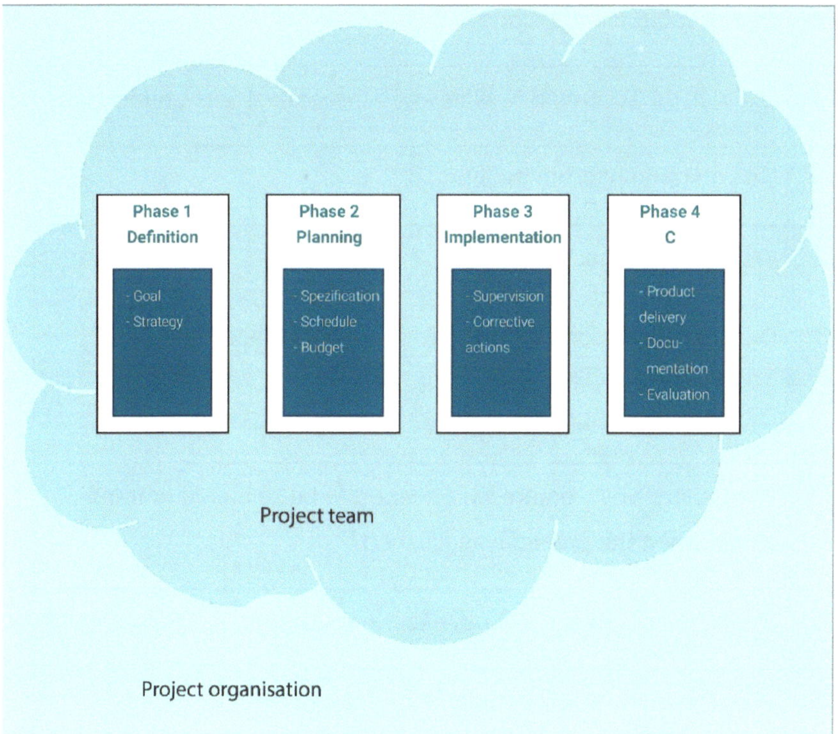

Figure 2.2: The four project phases

Table 2.1: Answer the Following Questions Regarding your Own Project

1	When did you first get the idea for this project?
1.1	How much time passed and how many steps did you take between the first idea and a clear concept?
2	How did you start planning?

Continuation Table 2.1

2.1	Decide which tool, which equipment and materials you would need and where to buy them from?
2.2	Did you plan additional help in case the project would not be

	manageable for you alone?
3	Did it all go according to plan when the project was underway?
3.1	Do you remain within the budget?
3.2	Were you finished in time?
3.3	Did the results correspond to your quality standards?
3.4	Did unexpected problems arise? If so, how did you solve it?
4	Were there any workers who needed to be released or transferred when the project was finished?
4.1	Did tools and equipment have to be returned?
4.2	Was there any surplus material to dispose of?
5	Did you think about the experiences after the project was completed in order to find ways to improve the next project? If not, take a few minutes now and write down some suggestions for improvement.

Remember a project you have completed in the past two or three months. It could have been a weekend project at home or a professional project.

Table 2.2 discusses a bathroom conversion that was carried out without prior planning.

Table 2.2: Project Example "Bathroom Renovation"; course of a completed project

1	When did you first get the idea for this	About 3 months before.

	project?	
1.1	How much time passed and how many steps did you take between the first idea and a clear concept?	Very little.
2	How did you start planning?	Measure the space, draw a drawing, select sanitary facilities according to leaflets, enter dimensions, draw up a sketch of the objects, make a sketch of the position of the lamps, calculate square metres for tiles according to their selection, place orders, if necessary.
2.1	Decide which tool, which equipment and materials you would need and where to buy them from?	General pipe materials were procured by the craftsmen themselves, the bathroom equipment components were partly ordered too late, as the delivery times had not been known before.

Continuation Table 2.2

2.2	Did you plan additional help in case the project would not be manageable for you alone?	Yes, two craftsmen of an entrepreneur should carry out the construction work.
3	Did it all go according to plan when the project was underway?	No, some bathroom fittings were ordered too late and had a delivery time of about 6 weeks. In addition, special designs (shower tray) were added, as other di-

		mensions and situations arose after the eruption.
3.1	Do you remain within the budget?	Yes, approx. 5% more.
3.2	Were you finished in time?	No, the project had to be interrupted for two weeks because important components were missing.
3.3	Did you meet your quality standards?	Yes.
3.4	Did unexpected problems arise? If so, how did you solve it?	The sewage system was not sufficient for the toilet facilities and had to be improved by a pump system. trap is vacuumed empty. It had to be corrected.

Continuation Table 2.2

4	Were there any workers who were to be released or transferred when the project was finished?	The workers were able to work seamlessly on other projects during the break and afterwards.
4.1	Did tools and equipment have to be returned?	The tools and devices were immediately used in other projects.
4.2	Was there any surplus material to dispose of?	Waste material was continuously disposed of Tiles were retained in case of damage during repairs. They're in the basement now.
5	Did you think about the experiences after the project was completed in order to find ways to improve the next project?	Yes, in essence, the order times of the sanitary facilities (sometimes up to 6 weeks) should have been taken into account in the project. In some cases, however, this is not possible in the renovation of old buildings, because some situations only become apparent after demolition and then have to be planned again: the shower tray was finally bricked up by hand, the shower cubicle was ordered, the pump system was only found after a long search for the existing toilet system and was accordingly somewhat more expensive.

Continuation Table 2.2

5.1	If not, why don't you take a few minutes to write down some suggestions for improvement?.	A sanitary technician would have been better able to foresee these measures. In this case, the problem was not serious, as the employees were able to work on a different task at short notice. This was deliberately planned after the finding that a pause would occur. The difficulties with the toilet facilities were not to be expected in advance. The originally intended installation would have been much more expensive and costly, so this was the better alternative. Luckily, there was a temporary accommodation for the remodelling weeks. If the apartment had to remain habitable, this construction measure would not have been feasible.

The final evaluation of the project is very important. This is described in more detail in chapter 7. From the mistakes made, it is only in this way that experience for future projects can be derived.

2.4 The Project Parameters

During the duration of the project, the project manager has to observe the three most important parameters:

- Quality
- Cost
- Time.

A successfully managed project will be completed on or before the due date, within budget and with the specified quality standards.

Each of these project parameters is precisely defined in the planning phase. These parameters then form the basis for monitoring the progress of the project during the implementation phase (table 2.3).

Agreement of Specifications

In order to be able to evaluate the quality of a project, specializations are created that describe the successful outcome of the project in detail. They are precisely coordinated with the client and included as part of the project contract. The client will check these characteristics at handover or acceptance of the project.

Such a client can be an external company or another department in your own company. The case study described here is a hall construction project designed to create more production space in the company. In this case, the customer is the department that wants to use the hall later. It sets the quality requirements. The management determines the cost and time frame.

Table 2.3: Projektparameter

Project Parameter		Basics for Monitoring the Project Progress
Quality	→	Spezifications
Cost	→	Budget
Time	→	Time schedule, dates

During the course of a project, the specifications may change. It is now up to the project manager to ensure that the client accepts the new specifications. If a written contract exists, it must be revised in writing so that there is no misunderstanding of the accepted parameters during the inspection/acceptance of the project result. At the same time, it is always clarified who will assume the costs that may arise.

Budget Preparation

The cost plan and thus the budget are derived from the available financial resources. If there are not enough funds available, the quality requirements may be changed or internal services may be provided.

Drawing up a Schedule and the Corresponding Dates

The scheduled time of completion of the project will determine the schedule and thus the dates. If there is not enough time available, the quality requirements are changed or additional services are purchased.

Today and in the future, however, it is not only these three parameters that determine the project decisively. There are three more secondary parameters

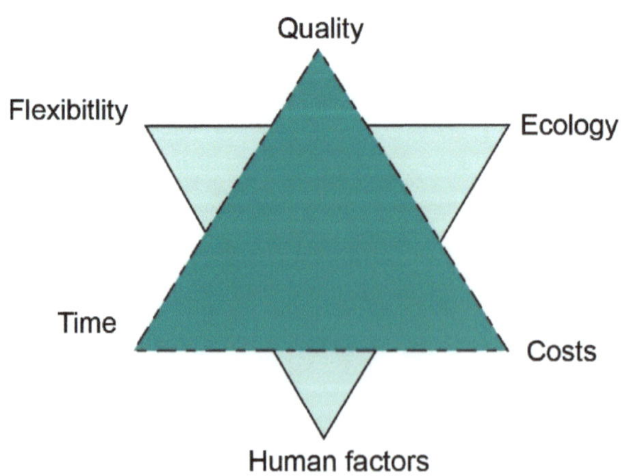

Figure 2.3: Targets for Optimizing Development Projects

- Ecology
- Flexibility
- Human Factors

together with the primary parameters in figure 2.3.

These parameters are particularly important for development projects.

Ecological awareness and the resulting progressive legislation lead to increased demands with regard to the environmental compatibility of projects.

These parameters are particularly important for development projects.

Ecological awareness and the resulting progressive legislation lead to increased demands with regard to the environmental compatibility of projects.

Today, the company must be able to react flexibly to every customer request. This is the only way to cope with the changed market situation.

Only the consideration of human factors will achieve the necessary motivation and identification, for example with the company, its organizational structure and the project to be carried out.

Table 2.4 lists the skills required by a successful project manager. Mark all the skills you think you have. Then mark the skills you want to improve in the second row. Consider which skills are especially needed in your projects.

2.5 Making the Right Projects

In project management, too, marketing is essential. Convincing superiors and employees of plans and gaining their support is not possible without preliminary work. The best project handling, i. e. to do the project correctly, is of no use if the wrong project has been selected.

Table 2.4: Skills Required by the Project Manager

1	Organize projects from start to finish.	☐	☐
2	Structuring plans that withstand project pressure.	☐	☐
3	To convince supervisors and employees of the project plans and to gain their support.	☐	☐
4	Clearly assessable project goals.	☐	☐
5	To lead employees.	☐	☐
6	Support employees with problems.	☐	☐
7	Planning and optimal use of existing resources	☐	☐
8	Eliminate waste of time and money.	☐	☐

Continuation Table 2.4

9	Measure the progress and performance of a project.	☐	☐
10	Appropriate use of information systems.	☐	☐
11	Identify and resolve conflicts at an early stage.	☐	☐
12	Define project goals comprehensively.	☐	☐

The right project is also the right one if the available resources, such as personnel, material or financial means, can be used correctly in order to realise the proposed project. Even an interesting project can fail due to a shortage of resources.

The requirement is to realize the goals of the company best with the selected project. However, the fulfilment of this demand is not always easy.

In fact, this systematic selection often fails because of the difficulties in the organisation, for example, because of the so-called hereditary farms. The projects are often selected by feeling. The prestige of the applicant, the orientation towards already known costs incurred for similar projects, such as budget quotas of the past, are frequently used but totally inadequate selection procedures. Current market-decisive factors must be taken into account.

A simple method is to determine the success factors of a project and the project profile derived from it. Success factors are the characteristics of a product or the organization offering the products, which significantly determine the competitive position of the company on the market. The biggest advantage of such an orientation towards these success factors is that the product development receives an important support in formulating the project goal. In addition, the procedure also takes into account important factors that are decisive for purchasing decisions.

The compilation of the project's success factors, for example a product development, is compared as far as possible with those of the competition. This means that you know your own market advantage and market disadvantage when this development project is started. Table 2.5 shows such success factors and levels of fulfillment of a product in comparison with the strongest competitor.

This overview shows that the planned development project has significant market disadvantages compared to the strongest competitor in several important aspects. Only when it has been clarified whether all these disadvantages can be compensated for, can a successful project be expected.

Table 2.5: Success Factors and Levels of Fulfillment of a Product Development Compared to the Strongest Competitor; ◆◆◆◆ Own Company; ◎◎◎◎ Strongest Competitor

	Success Factors	fulfillment levels bad good
1	Fulfillment of the specified product functions-	◆◆◆◆◆◆◆◆ ◎◎◎◎◎◎◎
2	Good product design-	◆◆◆◆◆◆◆◆ ◎◎◎◎◎◎◎
3	Customer group-specific product differentiation.	◆◆◆◆ ◎◎◎◎◎◎◎

Continuation Table 2.5

4	Consulting competence of the sales department or seller.	◆◆ ◎◎◎◎◎◎◎◎◎◎◎ ◎◎◎
5	Support of the customer in planning and installation.	◆◆◆◆◆◆◆◆◆◆◆ ◆◆ ◎◎◎◎◎◎◎◎◎◎◎ ◎◎◎
6	Use of system specialists.	◆◆◆◆ ◎◎◎◎◎◎◎
7	Fulfillment of special requirements.	◆◆◆◆ ◎◎◎◎◎◎◎

8	Good functional description.	◆◆◆◆ ◎◎◎◎◎◎◎
9	High acceptance of the products by the user/consumer.	◆◆ ◎◎◎◎◎◎◎◎◎◎◎ ◎◎
10	Brand name.	◆◆◆◆◆◆◆◆◆ ◎◎◎◎◎◎◎◎◎◎◎ ◎◎
11	Brand awareness.	◆◆◆◆ ◎◎◎◎◎◎◎

Continuation Table 2.5

12	High importance of market leadership.	◆◆◆◆◆◆◆◆◆ ◎◎◎◎◎◎◎◎◎◎◎ ◎◎
13	Own customer network.	◆◆◆◆ ◎◎◎◎◎◎◎
14	Focus on key accounts.	◆◆◆◆ ◎◎◎◎◎◎◎
15	Low susceptibility to faults.	◆◆◆◆ ◎◎◎◎◎◎◎
16	State-of-the-art product.	◆◆◆◆◆◆◆◆◆ ◎◎◎◎◎◎◎◎◎◎◎ ◎◎

17	Wide range of functions.	◆◆◆◆◆◆◆◆ ◎◎◎◎◎◎◎
18	Satisfying security needs.	◆◆◆ ◎◎◎◎◎◎◎
19	Joint demand with other products.	◆◆◆◆ ◎◎◎◎◎◎◎
20	Easy maintenance.	◆◆ ◎◎◎◎◎◎◎

2.5.1 Case Study Car Tires

Figure 2.4 shows the factors that determine the purchase of car tyres. The most important characteristic in the buyer's judgement is the driving behaviour. The feature profile shown was created using a simplified marketing method /2.3/.

Therefore, the property that contains the buying factor is broken down in more detail (figure 2.5). Here, the all-year-round and wetting suitability as a product feature that is highly valued by the customer is decisive.

Based on these survey data, which are assigned to a product and which can be designed in a development project, the product profile is constructed for evaluation (table 2.6).

The various properties are compared with those of a particularly successful competitor's product. Project 1, which concentrates primarily on significantly improving the customer's estimated driving behaviour, should be preferred to Project 2.

2.5.2 Decision-Making Rulse

The following decision-making rules for the realization of a project can be established:

- With the significant success factor "price of the product", manufacturing and subsequent maintenance costs can be reduced.

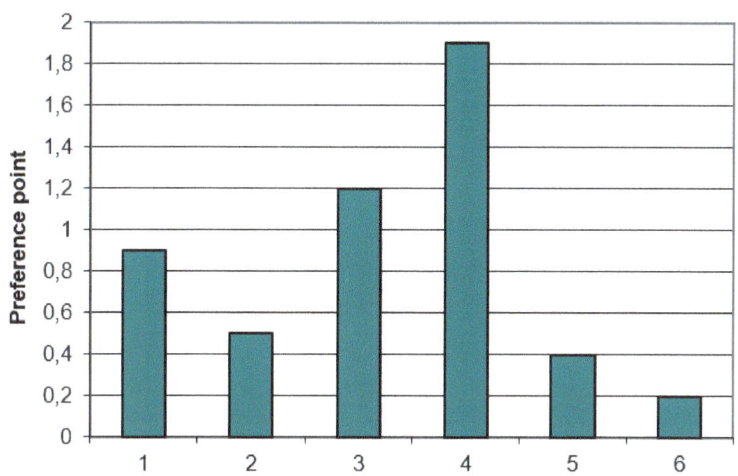

Figure 2.4: Success factors for the purchase of car tyres; 1 driving characteristics; 2 driving options; 3 price; 4 driving characteristics; 5 brand preference; 6 optics

- In the case of success factors "customer remuneration", the mandatory targets, i. e. the goals that must be achieved without fail, are to be assigned to the most important success factors. The strongest competitor in terms of performance parameters is surpassed or residues can be made up for.

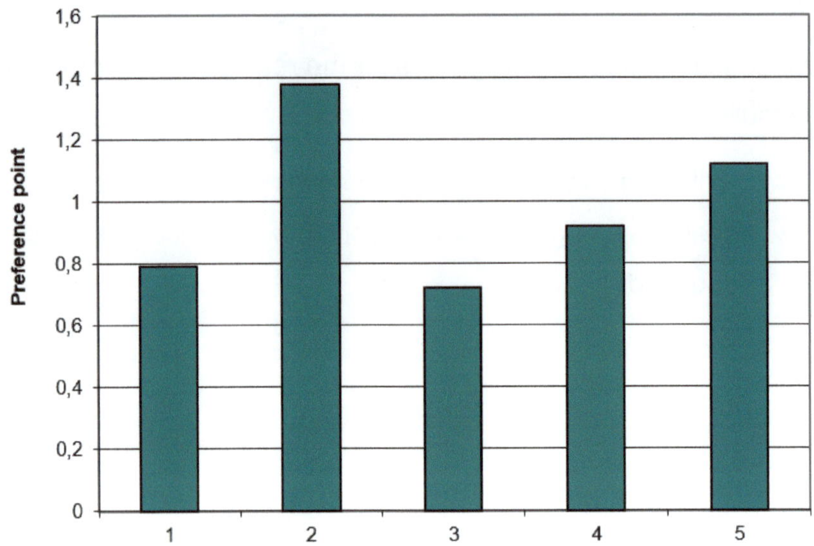

Figure 2.5: Decisive factors for driving behaviour; 1 behaviour; 2 wet suitability; 3 suspension comfort; 4 service life; 5 all-year suitability

- However, commercial aspects must also be taken into account alongside other criteria, such as
- o the expected return on investment, i. e. the profit related to the investment
- o the expected return on sales
- o the forecast contribution margins.

The investor has several methods at his disposal for selecting the right project, but these are not explained in detail here:

- the cost comparison calculation
- the profitability calculation
- the amortization calculation
- the capital value method
- the internal interest rate method

- the annuity method.

Table 2.6: Product Specification of Two Different Planned Projects for the Development of Car Tires

Comparison with Competitors				
Success Factors	**Fack-tor**	**Weighting** worser		better
Driving characteristics.	0,18	1	2	
Driving option.	0,12	2		1
Prize.	0,19		2	1
Driving behaviour.	0,40		1	2
Brand preference.	0,09		1, 2	
Optics.	0,02	1	2	

The contractor who will carry out the project also has to ask himself whether he will take on the right project or whether he should even try to get the contract for the project, as the tender costs can amount to up to 5% of the contract value.

table 2.6 and table 2.7, which are very similar to a risk assessment, are suitable for assessing this question.

The main criteria are capacity load, risks, competitive opportunities and customer characteristics, such as creditworthiness. These criteria are further sub-criteria.

The project receives a grade in each subcriterion. The scale ranges from 1, very poor compared to the average, to 5, significantly better than the average. The product of grade by weighting results in a score that is added to a total score. The example shows 369 points. This is compared with a score of 300 points (table 2.7 /2.3/) if all the criteria correspond to the average, i. e. grade 3. The evaluated project is therefore 23% above the average and has good conditions to be successful.

Afterwards, the project is assessed in the individual evaluation criteria in such a way that a decision can be made on the acceptance of the project. Table 2.8 clearly shows that the weighting of the individual criteria plays an important role in the project evaluation. For example, the project has a very favourable risk assessment and a lower customer specificity rating.

There are also very descriptive methods with the portfolio approach. Here, the projects to be selected are judged according to only two characteristics (dimensions)

- the economic and
- the strategic importance of the project.

Only projects in the black area (figure 2.6) have the highest priority and have to be selected because of their high economic and strategic importance. Projects that are located in the weakly marked areas have medium economic or medium strategic importance. All other projects within the white areas are either economically or strategically meaningless.

Nevertheless, there may be reasons to choose such a project. For example, if a new market needs to be conquered or your own resources need to be utilized.

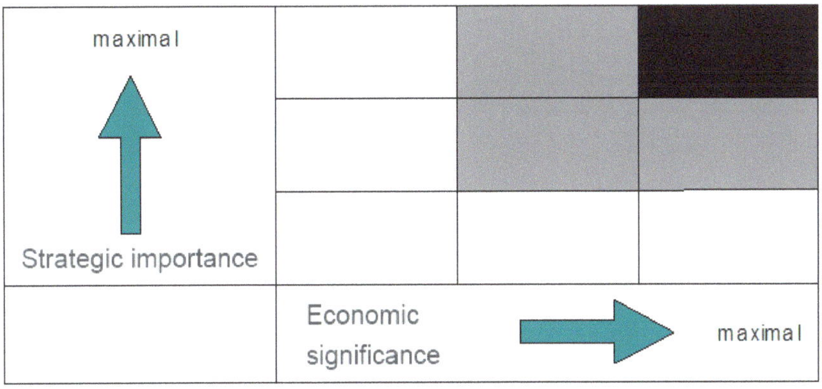

Figure 2.6: Priority of Projects According to the Portfolio Approach

Figure 2.7 shows the most important criteria for project evaluation before the start of the project. Usually, the company already has some experience, which is incorporated into the company.

2.6 Project Risks

In addition, the risks of the project that could endanger the fulfilment of the project goal must be assessed prior to the start of a project planning.

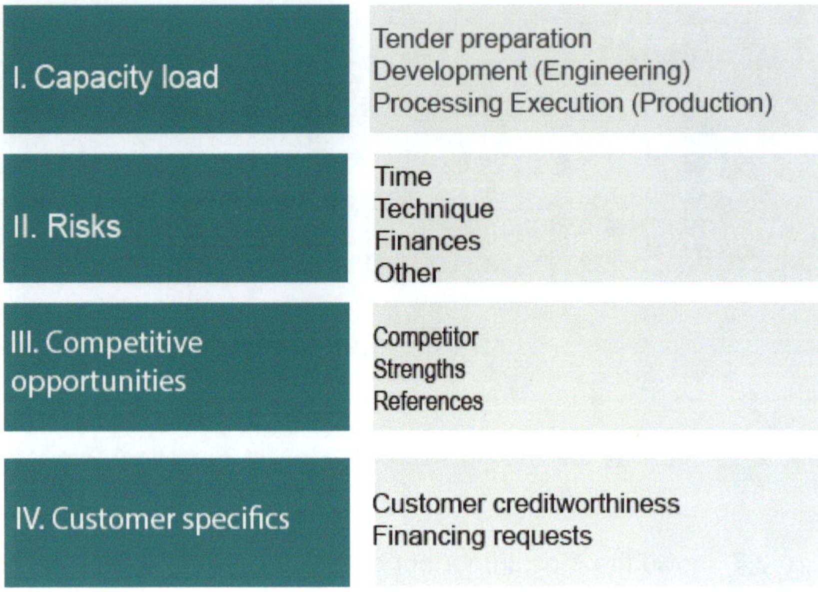

Figure 2.7: Criteria for Project Evaluation before the Start of the Project

Table 2.7: Project Evaluation before the Start of the Project /2.3/

Criteria for Project Evaluation	**Weighting %**	**Rating Grade Nmb.**	**Aver-age.**	**Diffe-rence %**
A Capacity Load				
A1 Proposal Preparation	4	16	12	33,3
A2 Processing (Engineering)	6	24	18	33,3
A3 Processing (Production)	10	50	30	66,7
Total Capacity Load	20	90	60	50,0
B Risks				

Continuation Table 2.7

B1 Time	6	30	18	66,7

B2 Technique	9	27	27	0,0
B3 Finances	6	24	18	33,3
B4 Other	9	27	27	0,0
Total Risks	30	108	90	20,0
C Competitive Opportunities				
C1 Offering Companies	12	24	36	33,3
C2 Strengths /References	12	48	36	33,3
C3 Order Opportunities	12	36	36	0,0
Total Competitive Opportunities	36	108	108	0,0
D Customer Specifics				
D1 Financing Requests	7	28	21	33,3
D2 Creditworthiness of the Customer	7	35	21	66,7
Total Customer Specifics	14	63	42	50,0
Total Evaluation	100	369	300	23,0

The list cannot, of course, be used to identify risks in detail. For this purpose, the details of the risks must then be specified for each area accordingly.

Table 2.8: Project Evaluation

Project Evaluation	**Grade**	**Weighting %**		**Valuation**

A Capacity Load	4,33	20	86,6	**************** *
B Risks	3,75	30	112,5	**************** ********
C Competitive Opportunities	3,00	36	108	**************** ****
D Customer Specifics	4,50	14	63	*************

The following questions help you to assess the risk assessment of personnel deployment:

- Are there any critical tasks for which no one has yet been designated?
- Are there any constraints to include certain employees in the project?
- Are there any constraints to employ more or less people in the early project phase?
- Do the members of the project team fit together?
- Do the members of the project team have realistic ideas about their tasks?
- Are the members of the project team really suitable for the tasks assigned to them?
- Are the most important members of the project team available for the expected project duration?
- Are the most important members of the project team completely assigned to the project?

Table 2.9, which should be filled out at the same time as the proposal is prepared, lists the possible risks of a development project.

Risks have to be thought through systematically, because many small corrections often lead to aggravations.

There are different types of risk

- the risk that is inevitable for all (weather, war),
- the risk we can't afford,
- the risk we can afford.

The risk must therefore always be evaluated specifically. If the company were to become insolvent when the risk materializes, it would not be able to enter into the transaction, for example.

Table 2.9: Risk Assessment of a Project

Potential risk factors	**Risk Elimination Measures**
Staff shortage/lack of adequately qualified personnel	Purchasing of specialists Team building Workout Long-term personnel planning, especially of key personnel

Continuation Table 2.9

Unrealistic schedules and budgets	Detailed cost and time planning Avoidance of "luxury functions Phased development (phases/versions) Targeted use of existing, possibly also imperfect resources.

Development errors	Analysis of the organization Creation of a prototype Use of simulation software Integration of the potential user Early creation of functional specifications and user manuals.
Development of faulty interfaces of the products	Creation of a prototype Use of simulation software Analysis of tasks Integration of the potential user
Equipping the product with superfluous functions	Creation of a prototype Use of simulation software Integration of the potential user Cost-benefit analyses
Continuous changes in requirements by the client	Contractual specification Consistent change management

Continuation Table 2.9

Defects in externally delivered or allocated components	Obtaining references Fixed price contracts Consistent entry control Integration of suppliers into the team
Expectations of the possibilities are too high	feasibility studies Creation of a prototype

	Simulation with the help of the prototype Use of simulation software

For this purpose, it is necessary to determine the boundary conditions, such as the price of gold or similarly relevant information for the project. However, this must be recorded in writing in order to become the yardstick.

Table 2.10 shows a detailed risk checklist for plant construction. Each industry will develop specific checklists to minimize the risk in the different project phases. The table is started from the offer phase and will be updated during the course of the project.

Table 2.10: Example of a Detailed Risk Checklist for Plant Construction

Risks from Dependencies	**Character of the Risk**		
	high medium low		
Dependence on suppliers			
Do low-cost companies or local companies* have to be used?			

Continuation Table 2.10

Does this lead to increased engineering and monitoring efforts?			
Are these "cheap solutions" still economical considering the additional costs?			
Does the customer prescribe suppliers?			
Does the customer have an extensive list of			

suppliers?			
Do suppliers come from other countries, unlike planned, for example local suppliers*? Does this have consequences?			
Are there any additional language problems or similar by local suppliers?			
Is there a strong dependence on subcontractors for engineering or assembly services?			
Can subcontractors be expected to fail to meet deadlines, deliver quality and reliability? Does this have negative effects?			

* for example, third world countries as clients

In the course of a project, new, as yet unknown risks can arise again and again, or already identified risks must be assessed differently. Therefore, risk analyses are carried out again and again during the implementation of the project.

Here you can book a free Strategy Session or write me if you like this book and have suggestions or questions.

Click here for free bonus material on the book.

Also visit my blog „Selbstführung & Produktivität in German“. I'll help you get better results.

3 Project Definition

Searching for destinations
Find ways
Taking small stepshen

Tripping Hazard

If there is no project definition at all or only an insufficient one!

In order to analyze the project's purpose and goal precisely, the project idea must be comprehensively illuminated from all sides. The more open questions can be clarified at an early stage, the more successful the project can be.

Tripping Hazard

When schedules and resources are incomplete or insufficient!

- Important activities are lacking.
- The time estimation for individual tasks is unrealistic.
- The dependencies between the activities are not taken into account.
- The required skills are not identified.
- Working hours are not detailed enough

The project definition has to be carried out again to clarify the missing points. There may be different project visions that need to be brought together.

Procedure for Project Definition

- Study, discuss and analyse,
- Define the project provisionally,

- List needs and wishes,
- Setiing a final goal,
- Develop alternative strategies,
- Evaluate alternatives,
- Choose a procedure.

This process may need to be repeated several times.

3.1 Objectives of the Project Definition

Projects can only be worked on and followed up in concrete terms if it is clear and known what is desired.

In order to define a project correctly, the project goal must be known as comprehensively as possible. It is always necessary to define a project goal as precisely as possible, because the motto is:"Who does not know the goal cannot find the way.

Only the clarity of purpose shows whether what is done in the project is relevant for the project goal and thus the success of the project.

For example, there are questioning techniques that support a methodical approach for exact target determination. These are described in detail in chapter 9.

3.2 The Origin of Projects

Projects are often the result of ideas that aren't very concrete, which come to mind on the side:"One should really...","That would have to be investigated...". You say yes quickly and you are enthusiastic, but you don´t see the mountain of work coming towards you that you hadn't expected.

That is why we should not first of all examine these ideas for their scope and substance. The individual steps are defined as precisely as possible.

A good method is to answer the w-questions: who, what, when, where, how, why?

Projects often arise from problems or opportunities. In their professional life they can be initiated by the company management, customers or employees, while in their studies they are taught by lecturers, students or the administration. A project is "born". If someone reacts to a problem or wants to start a new company and the decision is made to tackle the problem or the company, a project exists.

As a rule, someone is given responsibility for this: the project manager.

Problems can arise in every project, which have to be handled purposefully. The originator of a project is often not fully aware of important aspects of the project. The project team members also tend to overemphasize personal points of view during the definition and structuring stages of the project.

If such personal prejudices and conflicts of interest are disregarded, the catastrophic consequences for the project can be disastrous.

However, these aspects can be clarified through discussions and exchanges of views between the project manager, the client and the staff. By understanding the expectations, the project manager is able to begin with the definition of the project.

In large companies, where there are separate departments for the preparation of tenders, they often take over this part completely and hand over the project to the project manager only when the order is placed. It would be better to involve the project manager at an early stage, also in this phase of the project.

3.3 The Start of a Project

As soon as the core of the project team has been determined, the first task to be completed is to clarify the project. On the one hand, a consensus is reached among employees on the definition and scope of the project, and on the other hand, on the fundamental strategy of the implementation. The following seven steps lead to a smooth start of the project. They can also be run through several times, for example, if the result is still unsatisfactory after the first run.

Study, Discuss and Analyze

It is very important that the team has enough time to study, discuss and analyse the project, as this gives everyone a clear idea of the problem. It may be necessary to investigate the approaches in other projects, or to explore how other patterns of experience could contribute to project planning.

The purpose of this activity is to clarify that the right problem or opportunity is being considered.

Define the Project

If the project manager thinks he has the situation under control, he or she creates a preliminary project definition. This is, of course, constantly revised with the arrival of new information and the newly gained experience.

Setting a Final Goal

Either the final goal for the project must be determined by the project end finition itself or the final goal is determined by the client from the outset. The feasibility is then checked using the project definition.

List Needs and Wishes

First, a list of the absolute needs (need to have) is drawn up, which must be implemented in the final result of the successfully completed project. Afterwards, a second list can list the additional requests (nice to have) whose realization is not necessarily necessary for the successful completion of the project, but would increase it.

Developing Alternative Strategies

Alternative strategies are now being developed for the solution strategy, which could also lead to the goal. To find these alternatives, it is recommended to use creativity methods such as brainstorming sessions or creating mind maps to leverage the creative potential of the project team (chapter 9).

Assessing Alternatives

The evaluation of the alternative strategies found and developed in this way is the next step. It must be ensured that the chosen evaluation criteria, for example a marketing study or a risk analysis, are realistic and reflect the final goal.

Selecting a Procedure

This evaluation now allows you to select a more objective approach that best suits the project definition and the final goal.

3.4 Feasibility Study

Before starting a large-scale project, a feasibility study should be carried out to review the preliminary strategy and, of course, also to answer the fundamental question:"Will it work? to answer your question. Depending on the type of project, one of the following three alternatives can be considered: carrying out a market study, a pilot test or a computer simulation.

The funds invested in the feasibility study must, of course, be proportionate to the costs of the project. For example, for a company planning to build a new production facility for €200 million, a market study for €200,000 is an appropriate investment to assess the project risk.

At the other end of the scale, for example, there is a butcher's shop chain with a small regional network of branches that is planning to introduce a new sausage speciality. The market study for this product is limited to a test involving only a few stores. The sales figures for one week alone are a good indication of the market potential of this new delicacy. The relatively low investment in local advertising and the production itself provides the desired information.

3.4.1 Market Study

If the project is to bring a new product onto the market, its market potential must be researched. Market researchers ask prospective customers whether the product meets their actual or potential perceived needs. In addition, similar products that are already on the market can also be investigated in order to draw conclusions about the chances of the planned product.

3.4.2 Pilot Test or Prototype

A pilot test or the construction of a prototype is a small trial run of the project. This can be a limited market test or a 1:100 or even a 1:1 model of the product. These tests, which are sometimes very time-consuming, offer the possibility of analysing the project results under realistic conditions.

This is often unavoidable for new technical developments.

3.4.3 Computer Simulation

Thanks to today's technologies, many projects can be virtually represented as computer models. For example, the behavior of a product can be scrutinized on the basis of the analysis.

The static and dynamic behaviour of buildings or machines is determined by mathematical and numerical calculations. This computer simulation is used in such diverse fields as aerodynamics and thermodynamics or optical and mechanical design. In many cases, the computer nowadays assists in the actual design of the product.

The main purpose of a simulation is to identify potential problems before the construction of a prototype and not just the prototype itself. This will then be almost perfect and will have no significant changes.

The development times of new developments are drastically reduced.

3.4.4 Utilizing Study Results

If the result of a well thought-out and executed feasibility study is good, the detailed planning and implementation of the project can begin.

If the result is discouraging, the data obtained will be used for a revision of the project, followed by another feasibility study. Even if this process has to be repeated several times, this is the only way to develop a promising project concept.

3.5 Test Your Progress

At this point you should pause for a moment and use table 3.1 to recapitulate the knowledge you have gained so far about projects.

Table 3.1: Chapter 3 Test: Which statements are correct or false?

No.	Question	correct	false
1	A project is a continuous or unlimited undertaking.	☐	☐
2	Quality is not important in projects.	☐	☐

3	The timely completion of the project is an important parameter of project management.	☐	☐
4	The project team must clearly define the project before starting.	☐	☐
5	Brainstorming is irrelevant for project management.	☐	☐
6	It is not important to adhere to the given budget framework.	☐	☐
7	The basic strategy must be reviewed before the project is carried out.	☐	☐
8	A pilot test can be used to check the strategy.	☐	☐
9	Computer simulations can prove the feasibility of projects.	☐	☐

Continuation Table 3.1

10	Project management is hardly different from other management applications.	☐	☐

1	2	3	4	5	6	7	8	9	10
f	f	r	r	f	f	r	r	r	f

Task

- Display the characteristics of a project.
- Name typical project phases.
- Which measures can make a project more successful?
- Formulate five questions to help you assess the correct deployment of personnel.

4 Project Planning

It is about thinking itself, and not so much about the result of the reflection process.

Tripping Hazard

If the key information is not specified in writing!

- In a project, a lot of information is gathered that has to be logged, because no one can keep it all in mind. In addition, written agreements are more binding than verbal commitments.
- In the Job Procedure Handbook, all processes, responsibilities and key information are recorded in writing.

Planning is crucial for project management. Everything that is necessary for the successful completion of the project in terms of quality, time and costs must be specified in detail. Each of these three dimensions is dealt with below with some tools and techniques.

4.1 Planning Steps

This includes the following planning steps:

- Set the goal of the project.
- Select a basic strategy to achieve the goal.
- Divide the project into individual steps.
- Define the performance criteria for each individual step.
- Set the time required for each individual step.
- Set the correcte order for the completion of the individual steps and transfer this information to a schedule for the entire project.
- Calculate the costs for the individual steps and determine the budget of the project.

- Define the personnel structure plan with the corresponding job descriptions.
- Define any necessary employee training courses for the project team.
- Developing the necessary guidelines and procedures (Job Procedure Book).

It can happen that even these steps have to be repeated if it turns out at the end of the planning that one or more project dimensions are exceeded, for example the total cost of the project, the budget.

At the beginning of the project, important dates and prices may still be missing. Then they must be estimated as realistically as possible. The correct data is then inserted into the planning later. For this, the project manager needs experience: his experiences, those of specialists and those from previous projects. These experiences are summarized in the project conclusion (chapter 7).

4.2 Planning the Three Project Parameters

4.2.1 Planning the Quality Dimension

In quality planning, all details of the project must be observed, because the aim of this planning is to guarantee the performance and functionality of the project result and to be able to objectively evaluate it.

Tripping Hazard

If the goals are too vague!

In this case, the project definition has to be worked through again in order to clarify the objectives.

Creativity techniques such as brainstorming, mind mapping or research can help.

A precise definition of the project's objectives is used to determine why and how the process is to be carried out.

For example, the description "red paint" is not sufficient. It is described exactly with which colour, for example the RAL number, which quality, which colour application thickness, etc. must be produced.

The quality plan also defines the performance characteristics and criteria according to which the project result is finally evaluated after completion.

Three things are needed to plan the quality dimension:

- the specifications for the materials used (quality, grade),
- the performance standards to be met, and
- the procedures to be followed for quality control and assurance.

Two techniques enable quality planning:

- a structure plan and
- the creation of project specifications.

In the technical department, specifications are often drawn up in the tender phase, which describe the quality and requirements of the project result.

4.2.1.1 Creating a structure plan

The structural plan is the starting point for planning the three project parameters quality, costs and time.

The division of a project into individual steps or work steps is based on this technique. Then all the elements necessary for the completion of the project are known. This minimizes the possibility of forgetting or neglecting important steps.

A structure plan divides the project into several levels. The more complex a project is, the deeper the structure.

Figure 4.1: Model Structure Plan

Logical subdivisions of the project are searched for and then broken down further. For example, a production plant is divided into its parts because the spatial overview is required for processing and especially for safety measures. The aim is to establish independent workflows that will bring the project closer to completion.

Figure 4.2 shows the structural plan of the construction project of a machine hall (hall construction project). The most important components of the construction project are subdivided into these. This makes the structure of the project visible.

This project is taken up in the following as an example project again and again. The project definition of the hall construction project is: Creation of production area by the construction of a machine hall in the period of 3 months and the costs of a maximum of EUR 350,000.

Task

Select the project from task in Chapter 2.3, which you have completed or plan to complete. Split this project into individual steps and then create a structure plan that shows the links between these individual steps.

But it is not only technical problems that can be better understood with the help of the structure plan. Figure 4.3 shows the structure plan of a Vietnamese wedding celebration, which is celebrated in a really big way. (The costs will be borne by the usual gifts of money of the guests.)

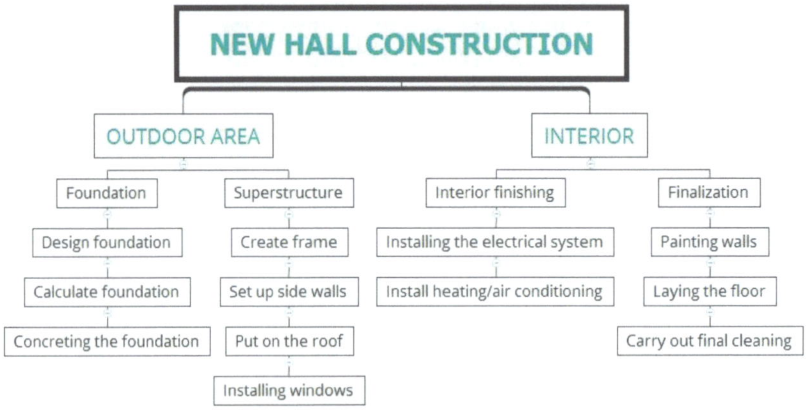

Figure 4.2: Case Study: Hall Construction Project; Structural Plan; Time: 3 Months; Costs EUR 350.000

4.2.1.2 Create Project Specifications

The project specifications (table 4.1) can now be created for all individual steps from the structure plan. These include all the relevant requirements to meet the quality dimension of the project. For example, these are the tools used, the performance criteria and the appropriate measures to verify these characteristics.

The preparation is carried out with great care, as it is the basis for performance monitoring during project implementation and has a direct impact on both schedule and budget.

Figure 4.3: Case Study: Vietnamese Wedding Celebration; Structural Plan; Time: 1 Year; Costs: EUR 30,000

4.2.2 Planning the Time Dimension

The time dimension of a project is investigated in order to find the shortest possible time schedule and timeframe for a project.

The structural plan is used to determine the necessary time for each individual step of the project, partly by estimation, if nothing else is known. The sequence of these single steps is then determined, or the individual steps that can run parallel to each other are selected.

Table 4.1: Case Study Hall Construction Project; Example of the Project Specification for the Foundations

Pour a 10 cm thick concrete slab over a compact, 15 cm thick sand filling. Reinforce these with reinforcement mats according to statics. A 0.6 mm thick polyethylene membrane as a water-repellent barrier be-

tween sand and concrete.
According to drawings, lay concrete beams? 30 x 45 cm around the edges of the foundation and underneath the load-bearing walls. Reinforce the beams with reinforcement steel? 6, which are placed with stirrups? 3 to 75 cm thick middle sections in each corner.
The concrete has a strength of 250 N/mm2 after 28 days.

Based on this analysis, the three most important time parameters are determined:

- the duration of each individual step,
- the earliest time for the beginning of a single step,
- the latest time for the start of a single step.

To plan the time dimension, you need to know the duration of the individual steps. If the project manager does not have sufficient expertise, he or she must work with a project team expert or an external specialist to determine the duration. If there is no other way, instead of specifying very exact values, a time span with a certain bandwidth is assumed first and then entered exactly when further information is available.

4.2.2.1 Time Control Models

As soon as the time period for each individual step of the project has been determined, the earliest and latest start time for each individual step can be determined.

In practice, two methods are used to graphically represent the project:

- GANTT charts /4.1/ and
- PERT diagrams.

4.2.2.2 GANTT-Charts

The GANTT chart displays the temporal links between the individual steps of a project as a horizontal bar chart. Each individual step is represented by a line, the length of the process is the duration of the process in which this single step is to be executed. The complete GANTT chart then shows the sequence of activities in a row, as well as those that can take place at the same time.

To create the chart, all individual steps are listed with the estimated duration required to complete the project. For this purpose, a bar is drawn at the height of the individual step for its duration (figure 4.4).

Some parallel single steps of different duration can be laid at the same time. This makes the individual step more flexible for the start time, as long as it is ensured that these individual steps are completed in time. Then they can be connected to the following steps. This results in time buffers, which are represented in the diagram as dashed lines. They are continued until the point in time at which the individual step must be completed to guarantee the overall schedule.

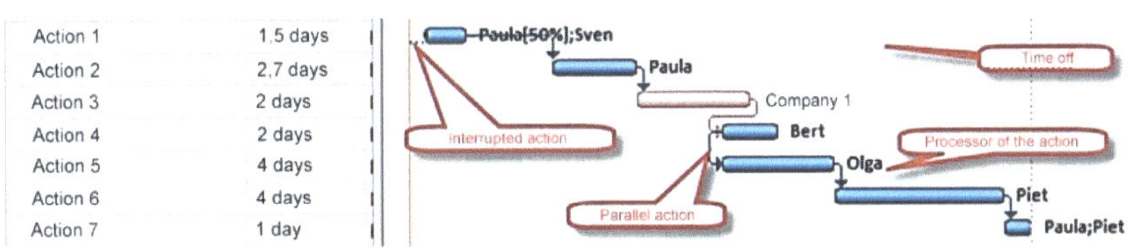

Figure 4.4: Example of a GANTT Chart

A GANTT chart shows that

- the minimum total duration of the project,
- the correcte sequence of the single steps and
- the simultaneity of individual steps.

However, the usefulness of a GANTT chart can be increased even further if, in addition to the planned data for the start and end of the individual steps of the project, the actual duration of the workflows is entered in a different color. This allows you to determine whether the entire project is running on schedule.

In the MS Project® project management program (chapter 8) a basic plan is created and can then be used in the monitoring mode. This requires additional time to enter the data.

In small projects, it is often sufficient to check or uncheck the single step in the plan.

Table 4.4: Case Study: Hall Construction Project; GANTT Chart; Project Steps with Time Estimation

No.	Project Step	Time[t]
1.	Draw up routings.	15
2.	Obtain building permission.	16
3.	Forming and concreting foundations.	5
4.	Setting up frames for walls and roofs.	5
5.	Cover roof.	5
6.	Install windows.	1
7.	Outdoor installation.	10
8.	Exterior painting.	3
9.	Laying the electrical network.	10
10.	Install heating/air conditioning system.	5
11.	Install insulation.	5

12.	Installing the interior panelling.	5
13.	Install interior doors.	5
14.	Interior painting.	3
15.	Installing electrical systems.	2
16.	Cleaning the building.	3
17.	Lay floors.	2

Figure 4.5 below shows the GANTT chart for the Indoor Construction Activity case study over the planned period.

The following table 4.4 describes the individual steps with the planned durations.

When describing these individual steps, it is important to describe the process as comprehensively and unambiguously as possible. This means in plain language: it must be called "install the windows", i. e. it must contain at least the subject and verb to actually express what is meant. The noun "window" alone will not suffice, because it could also mean "glazing the windows", for example, and the word "install" alone is not enough either, because something completely different could be installed, for example the heating.

4.2.2.3 PERT Diagramme

PERT is the abbreviation for Program Evaluation and Review Technique. It is a more sophisticated planning form than the one with GANTT charts and is particularly suitable for projects with many independent steps.

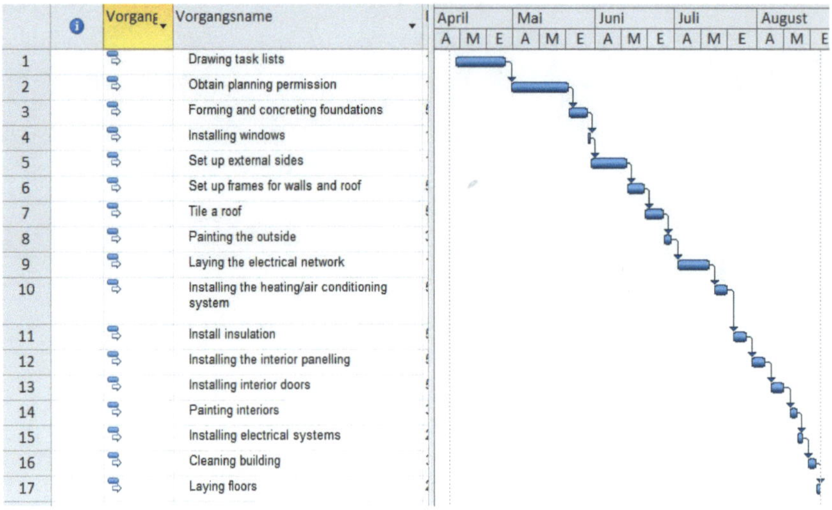

Figure 4.5: Case Study Hall Construction Project; GANTT Chart; Project Steps

The PERT diagram contains 3 components:

- Events are represented by event nodes, such as circles (figure 4.5) or other figures,
- Activities and their durations are represented by arrows between the event nodes,
- Dependencies that link two events, but for which there is no workload, are represented by dashed lines (figure 4.6).

PERT diagrams are the most common way to display the time planned to complete an activity. The duration is given in appropriate units corresponding to the size of the project, such as days, but also hours, weeks or months.

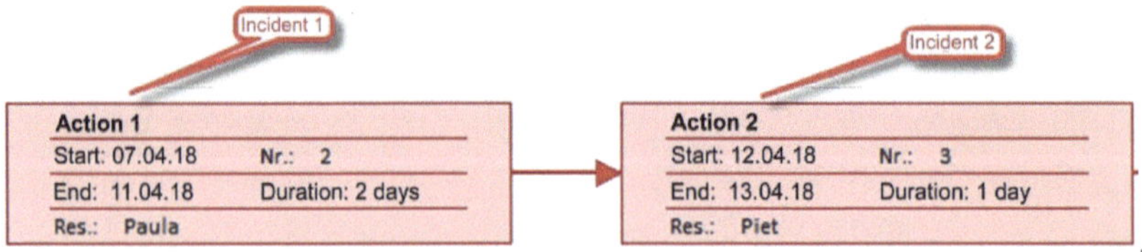

Figure 4.6: Principle of a PERT Diagram

The most mature PERT diagrams are drawn on the time scale, where the length of the arrows corresponds exactly to the duration of the activation. With a true to scale drawing, however, some arrows will become longer than they are actually needed for the job. These extensions represent the time intervals of the project. The arrows are marked with a clear dot at the end of the time period that is actually needed and are then dotted to the next event (figure 4.7).

To create a PERT diagram, all the individual steps required to complete the project are listed and their duration is also estimated in advance. Then a network of connections is drawn between these individual steps, whereby the sequence must be followed correctly. The numbers of the individual steps are transferred to the event nodes, and the time until the end of the following individual step is noted on the arrow. The individual steps that take place at the same time are entered on different paths. Finally, the system checks whether all individual steps are entered in the diagram.

A PERT diagram not only shows the relationships between different individual steps in the project, but it is also a way to calculate the critical path.

The critical path is the longest time path through the network of the PERT diagram and shows the most necessary work steps, the timely completion of which is essential for the progress of the project. If these times cannot be met, the project is automatically extended. In figure 4.8, the critical path is represented by the continuous arrow lines.

The PERT diagram is used as a control instrument during the project. The completed steps can be highlighted in color in the diagram, for example. The time required for completion is entered in color next to the planned one, in order to represent the actual progress of the project along the critical path.

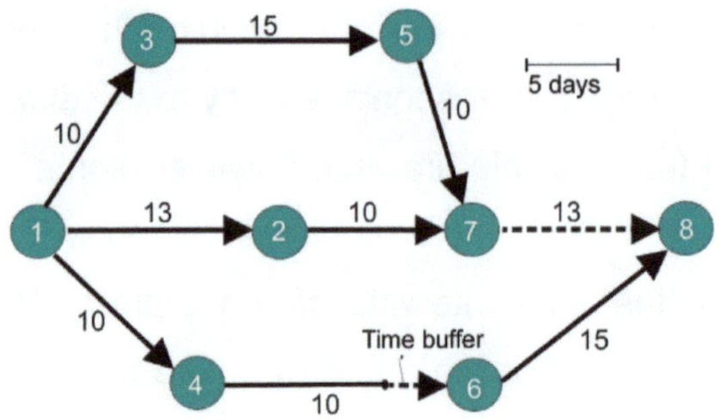

Figure 4.7: Elements of a PERT Diagram

Table 4.5 lists the project steps and their time data belonging to the PERT diagram in figure 4.7.

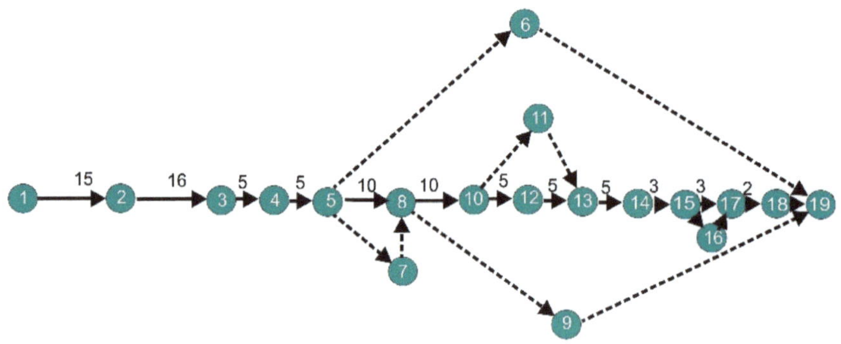

Figure 4.8: Case Study Hall Construction Project; PERT-Diagramm

Table 4.6 shows the individual steps for a finite element analysis (FEM) to calculate the eigenfrequencies of a component with MS Project®. In addition to the duration, you can also specify the interdependencies of the individual steps among each other (predecessor (VG), successor (NF)), the respective individual step costs and the respective project employees, including their percentage shares, with time and cost estimates, as well as the degree of completion.

To maintain costs, the resource plan includes the hourly rate, regular working hours, overtime pay, and so on for each project member.

Table 4.5: Case Study Hall Construction Project; Project steps with time Specifications

No.	Project Steps	Time[t]
1.	Project launch.	-
2.	Work schedules completed.	15
3.	Obtain building permission.	16
4.	Foundation concreted.	5

Continuation Table 4.5

No.	Project Steps	Time[t]
5.	Walls and roof frames erected.	5
6.	Roof completed.	5
7.	Window built in.	1
8.	Outsides set up.	10
9.	Exterior painted.	3
10.	Electrical network relocated.	10
11.	Heating/air conditioning system installed.	5
12.	Insulation completed.	5
13.	Interior panelling installed.	5
14.	Built-in interior doors.	5
15.	Interior painted.	3
16.	Electrical systems installed.	5

17.	Cleaning completed.	3
18.	Floors laid.	2
19.	Project completion	-

Table 4.6: Fallstudie: Finite-Elemente-Berechnung; GANTT-Chart

	Activity Name	Duration	Successor	Start Date	Final Deadline	Stage of Completion (%)	Predecessor	Costs (€)	Reviewer
1.	Creating a Model	4,77t		Fr 10.07.	Sa 18.07.	65		1.269	Ph;P
2.	Create CAE modelcorrections	0,43t	3	Fr 10.07.	Fr 10.07.	0		121	Ph[50%];P[50%];K[10%]
3.	Execute static run	1,07t	4	Fr 10.07.	Sa 11.07.	100	2	242	Ph[90%];K[10%]
4.	Check static run	0,5t	5; 6	So 12.07.	Mo 13.07.	0	3	125	K
5.	Incorporate corrections	0,5t		Mo 13.07.	Mo 13.07.	80	4	75	Ph
6.	Discussion with IT company	0,5t	7	13.07	13.07.	100	4	370	K
7.	Incorporate corrections	0,5t	9	14.07.	18.07.	60	6	75	Ph
8.	Perform natural frequency calculation	3,17t		15.07.	20.07.	53		538	PH;K

Continuation Table 4.6

9.	Perform natural frequency calculation	3t	10AA	15.07.	19.07.	70	7	748	Ph[80%];K[20%]
10.	Discussion with IT company	0,5t		20.07.	20.07.	0	9AA+1t	370	K
11.	Incorporate corrections	0,5t		20.07.	20.07.	0	9AA	75	Ph
12.	Develop result presentation	2t		20.07.	22.07.	0		1340	K
13.	Create report	1,33t	14	20.07.	21.07.	0		910	Ph[20%];K
14.	Submission	1t		21.07.	22.07.	0	13	75	K

The following GANTT chart (figure 4.9) shows the sequence of the individual steps as a bar chart. The completed or partially completed individual steps are shown in blue, the unprocessed individual steps or their parts are shown in red.

Figure 4.10 shows the PERT diagram MS Project®. However, it is not always possible to create it "fully automatically" from the GANTT chart at the push of a button.

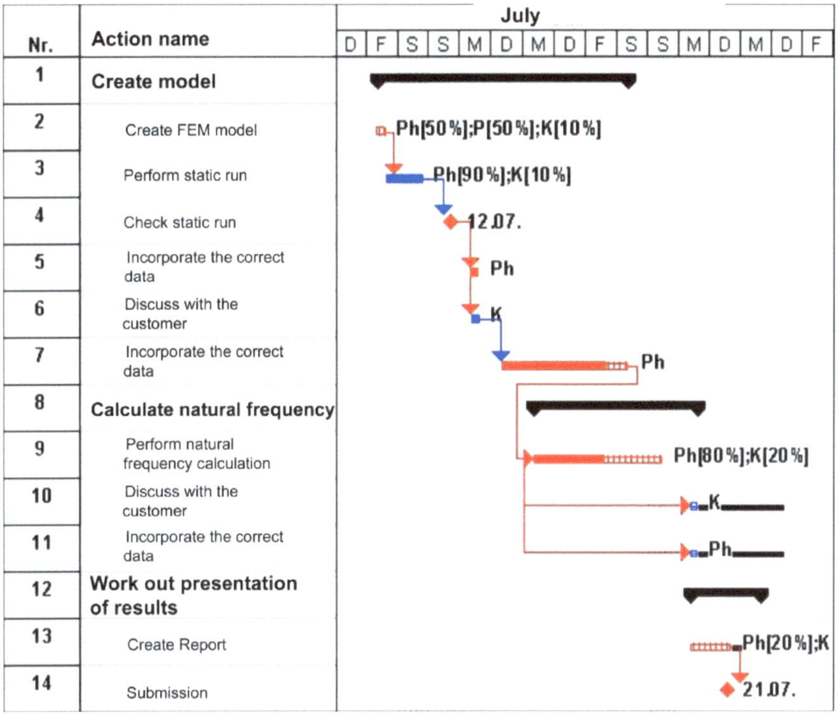

Figure 4.9: GANTT Chart in MS Project®

4.2.3 Planning the Cost Dimension

Today, it is essential to plan the project costs very careful and conscientious. If the costs are overestimated, the order could be lost because the offer is not competitive. However, if the costs are too low, you will notice in the course of the project that the project requires considerably more resources and you have to invest more to complete the project.

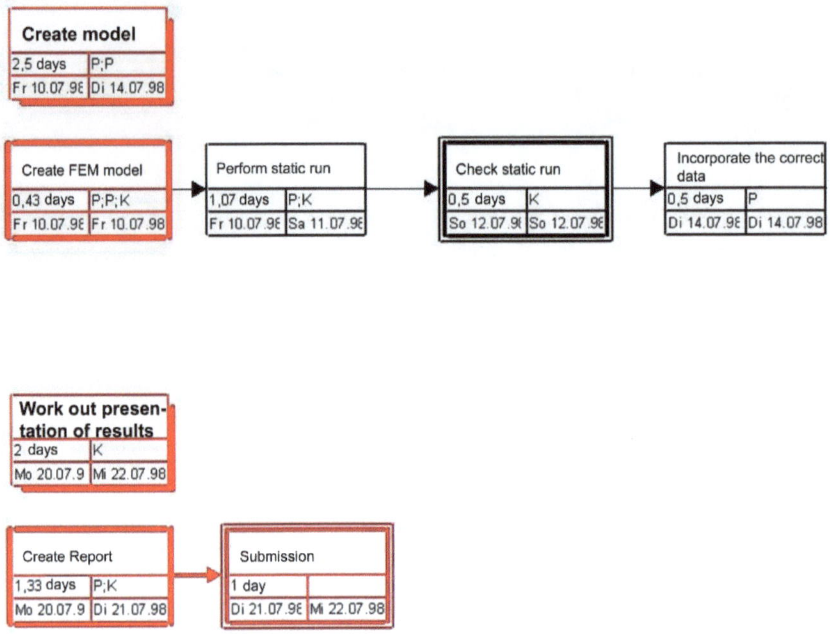

Figure 4.10: Section of a PERT Diagram in MS Project®

In order to obtain a good structural plan, the most cost-effective materials and equipment must be selected through a request on information, as well as other thorough research to realistically incorporate the actual costs incurred.

The project budget also has the function of monitoring expenditure during the course of the project to ensure that costs remain within budget.

Of course, not all inaccuracies cannot be ruled out when creating a preliminary budget, for example, due to the lack of knowledge of a single step when it is carried out for the first time. However, care should be taken to ensure that these inaccuracies do not arise from negligence in drawing up the budget. The cost planning should be as realistic as possible.

It is advisable to estimate the costs only when the time dimensions are known, since labour costs are often the most important cost category. The development of the project budget is thus based on the structure and schedule.

The individual cost categories are listed below:

- **Labour costs:** the wages and salaries of all workers and employees working directly on the project for the time spent on it
- **Overheads:** employer's contributions and fringe benefits for project staff (generally a percentage of total wage costs)
- **Material costs:** the cost of all items consumed in the project, such as timber, cement, cables, etc.
- **Equipment costs:** Costs of tools, equipment, office supplies, etc. required in the project; if a facility is used for the duration of the project, the costs may have to be apportioned over its duration
- **Equipment rentals:** the rental costs for large equipment, such as scaffolding materials, compressors, cranes
- **Ancillary costs:** the cost of management and non-supporting services such as purchasing, accounting, secretarial services, etc., for the time devoted to the project (generally a percentage of the total project costs)
- **Profit:** the profit from a successful completion of the project (generally a percentage of the total project costs)

Once the individual cost categories have been identified and the project is divided into individual steps, a work sheet is created to determine the costs for the entire project. Table 4.7 illustrates this for the case study: hall construction measures.

The costs of the individual steps, for example, subcontracting, are generally much easier to estimate than their own, because most of the orders are now placed at a fixed price. These costs then consist of the agreed price, the fixed price and the costs for the selection of the contractor and the award of the subproject.

Table 4.7: Case study hall construction project; project cost sheet in T€

Subunit	Labour	Overhead	Material	Equipment	Equipment Leasing	additional costs	Profit	Total Cost
1. Complete work plans	4	1,5		1		0,5	1	8
2. Obtain building permit						0,9		0,9
3. Concrete foundation	9	2,5	18	1,5	1,5	2,5	5	40
4. Put up walls and roof frame	13	3,5	23,5	2	9	2	7,5	60,5
5. Erect roof	7	2	9		1,5	1	2,5	23
6. Install window	7,5	2,5	23	1,5	5	3	5	47,5
7. Set up outer sides	12	2,5	30,5	1,4	9	2,5	7,5	65,4
8. Paint outer sides	3,2	0,8	0,5		1		0,5	6

Continuation Table 4.7

9. Lay the electrical net-work	5	1,5	3,5			0,5	1,5	12
10. Install hea-ting	5,5	1	21,5			1,5	3,5	32
11. Complete insulation	6	1,5	5				1,5	14
12. Install in-ner lining	5	1,5	5			0,5	0,5	12,5
13. Install inte-rior doors	3	1	7			0,5	1,5	13,
14. Painting of interiors	2,5	1	0,5		1		0,5	5,5
15. Install electrical equipment	1		1					2
16. Perform cleaning	2	0,5		0,5				3
17. Laying floors	2	0,5	3		0,5	1		7
Sum	**92,2**	**25,3**	**154**	**8,4**	**28,5**	**16,4**	**38**	**352,3**

4.3 Allocate Responsibility

Tripping Hazard

If roles and responsibilities are insufficiently defined!

- All responsibilities must also be recorded in writing.

Tripping Hazard

If not all persons important for the project are involved in the project!

- The most important project participants must be determined by the project manager and the names must be recorded in writing.

Tripping Hazard

If the team members are not made responsible for their work or do not feel responsible!

- Since there are no rewards or sanctions for fulfilling obligations in projects, the project participants must be motivated on their own initiative. This is best achieved through a clear project vision. The self-responsibility of the project team members can also be promoted through leadership techniques (e. g. the Instant Influence Method).
- The Instant Influence Method was developed by the American psychologist M. V. Pantalon /4.1/ has been developed. In only 6 steps you can pull them through in 7 minutes. It helps people to develop their own motivation. This is the only way to break down blockages and initiate sustainable changes in the shortest possible time.

It must be decided as early as possible who is responsible for completing each individual step of the project, so that those affected can participate in the planning of deadlines and budget. This participation naturally also increases the motivation and responsibility of the employees. If this is not done, a situation can arise in which no one feels responsible for the task (figure 4.11 and figure 4.12). Unfortunately, however, this is often the case in practice.

The number of people involved varies depending on the size of the project. In small projects, one employee can also be responsible for several individual steps or the entire project.

In order to make the best use of the resources available for assigning responsibilities, subcontractors and service departments should also be involved alongside the project team members.

In figure 4.13, the project manager is not assigned the full hierarchy (blue arrows) during maintenance work. The employees of the mechanical maintenance department do not report to him, but to their hierarchical superior. For example, he has to obtain information about the project from the head of mechanical maintenance, because he does not report to him in normal operation. Or they escape him completely.

Figure 4.11: Non-Distributed Responsibilities /Source: Unknown/

Only after the project manager has been assigned full responsibility during maintenance work (red arrow) does the work proceed correctly. This responsibility is given each time anew and only for maintenance work.

This is a story about four people
named Everybody, Somebody, Anybody and Nobody.
There was an important job to be done
and Everybody was asked to do it.
Everybody was sure Somebody would do it.

Anybody could have done it

but Nobody did it.

Somebody got angry about that

because it was Everybody´s job.

Everyone thought Anybody could do it

but Nobody realized

that Everybody wouldn´t do it.

It ended up

that Everybody blamed Somebody

when Nobody did

what Everybody could have done.

Figure 4.12: Non-Distributed Responsibilities /Source: Unknown/

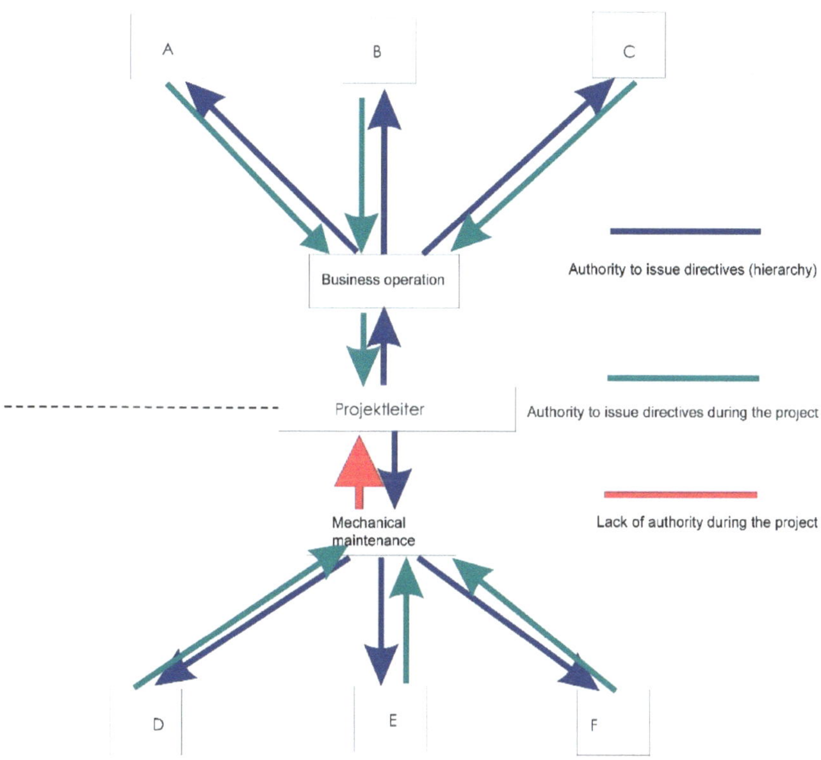

Figure 4.13: Assign Responsibilities Consistently a) without red: Influence Project Management, b) with red: Pure Project Management

4.4 Test Your Progress

With the questions in table 4.8 you can check your knowledge.

Table 4.8: Chapter 4 test: Which statements are correct or false?

No.	Question	correct	false
1	Planning is not necessary for small projects.	☐	☐
2	A structure plan is the starting point for project planning.	☐	☐
3	Working specifications illustrate the qualitative requirements of the project.	☐	☐
4	Tests should be part of the specifications.	☐	☐
5	When developing a project schedule, the duration and sequence of each stage is important.	☐	☐
6	The time at which a step can begin at the latest is not important for the planning process.	☐	☐
7	Experience is a basis for estimating the time required.	☐	☐
8	A GANTT chart illustrates the temporal relationships between the individual steps in a project.	☐	☐
9	A GANTT chart shows the temporal dependencies between project steps.	☐	☐
10	A PERT chart is more sophisticated than a GANTT chart.	☐	☐

Continuation Table 4.8

11	The critical path shows the most necessary work steps, the timely completion of which is essential for the progress of the project.	☐	☐
13	The persons responsible for subunits should not be integrated in the planning process.	☐	☐
14	The training of project staff is the responsibility of the human resources department and should not be taken care of by the project manager	☐	☐
15	Creating policies and assigning responsibility is part of the project process..	☐	☐

1	2	3	4	5	6	7	8	9	10	11	12	13	14	15
f	r	r	r	r	f	r	r	r	r	r	f	f	f	r

Task

- Show the planning steps of a project.
- Name the six cost categories to be considered for the project budget.
- Explain how to determine the time dimension of a project.

5 Project Organisation

Tripping Hazard

If management at the highest company level is inconsistent!

- The best way to achieve a project is to have the support of the company management. This requires a consistent perception of leadership responsibility at the highest level.
- Internal employees can rarely make a difference if they lack it. External, independent consultants are the better solution - if requested.

5.1 Basic Forms of Project Organization

A project organization is the configuration and composition of a project team, the organizational anchoring within the company, the function of the project organizator and the project manager, as well as the configuration of committees such as the steering committee.

The organizational implementation of project management within the company organization often leads to organizational instability of the hierarchy, because responsibilities have to be temporarily redistributed due to the necessary project organization. This always results in a reassessment of positions of power.

Today, project management is often used as a management tool. However, this often means for the project manager to lead without hierarchical competences.

There are three different forms of project organization:

- pure project organization
- influence project organization
- matrix project organization

Figure 5.1 shows the competence of the project manager within the respective organizational form.

5.1.1 Pure Project Organization

In this case, a project organization is created in parallel to the existing company organization. The employees required for the project are organizationally separated from the company's traditional departments and merged into a new, time-limited project organization. The project team founded in this way works exclusively for the objectives of the project. You will be led by a project manager who has extensive expertise and a high degree of responsibility.

The organisational form of pure project organization generally creates a streamlined form of work. The project manager has a large degree of autonomy and can react very quickly to faults. The employees identify with the project and thus contribute to the success of the project.

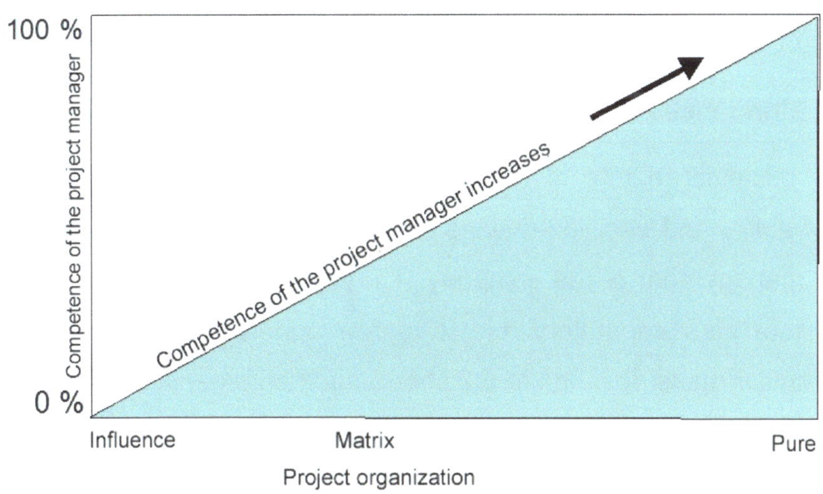

Figure 5.1: Competences of the Project Manager in the Respective Project Organization Forms

The real problem of this form of organization arises only when the project team is dissolved at the end of the project. This is all the more problematic the longer a project takes. New tasks have to be found for the members of the dissolved project team. Worst case, they have to be released.

Another problem with this form of organization is the high cost of the possibly multiple provision of resources in the form of material resources and personnel.

In addition, the competition between project organization and line organization can lead to a constant redistribution of power and, as a result, to unrest in the company.

Particular difficulties with this form of organisation are the recruitment of skilled employees, who are completely lost to the specialist departments as a result of the change in project organisation. In addition, they are further evaluated by the department for which they do not work.

For this reason, pure project organization is generally only used for large projects or project-leading departments, for example in plant engineering and construction.

5.1.2 Influence Project Organization

The essential feature of the project organization is that the full responsibility and decision-making authority is reserved to the existing line management of the company. The project manager has only an advisory and preparatory role, with few opportunities to implement unpopular measures. Figure 5.2 shows such an organizational form.

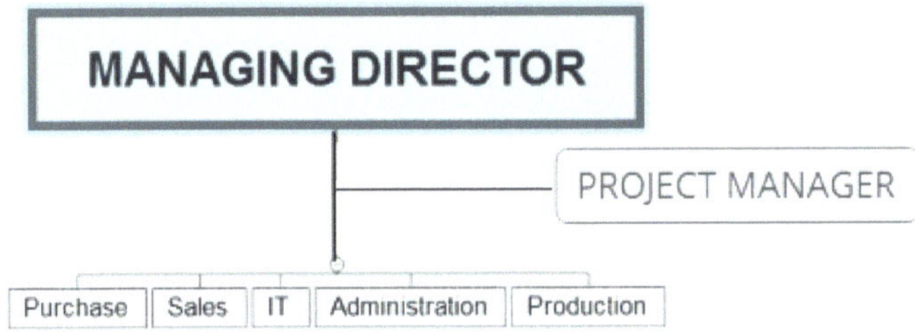

Figure 5.2: Influence Project Organization

The employees remain hierarchically in their organizational structures, which remain unchanged. The project manager has no special authority. He monitors the progress of the project in terms of technical, scheduling and costs and proposes the necessary measures. However, he should not be responsible for the project goal.

The project manager can thus only persuade the persons involved in the project to take the desired action, which he considers important for the progress of the project. Withdrawal of formal behaviour, for example insisting on agreements in the protocols, will not help him.

The organizational changes to be made are very limited, since the employees are only assigned for the respective assignment in the project, even on an hourly basis. The authority to issue directives and the decision on the further development of the employee remain in the specialist departments.

A reduction of the project manager to his or her role as a pure decision maker is rarely an advantage for the project flow. However, this organizational form has the advantage of high flexibility in terms of staff deployment.

However, the risk that no one identifies himself/herself sufficiently for the project goal is so great that this form of organization only makes sense for the smallest projects. The company management should always appoint a responsible project manager in the sense of a controller who controls and drives the project progress.

This is then the transition to matrix project organization.

5.1.3 Matrix Project Organization

Matrix project organization is a hybrid between pure project organization and influence project organization (figure 5.3 and figure 5.4). Here, the company's line organization is extended by an additional instance. There are project concerns that are assigned to the line offices and others that are transferred to the project team. This means that in a matrix project organization the power, but also the responsibility is divided accordingly..

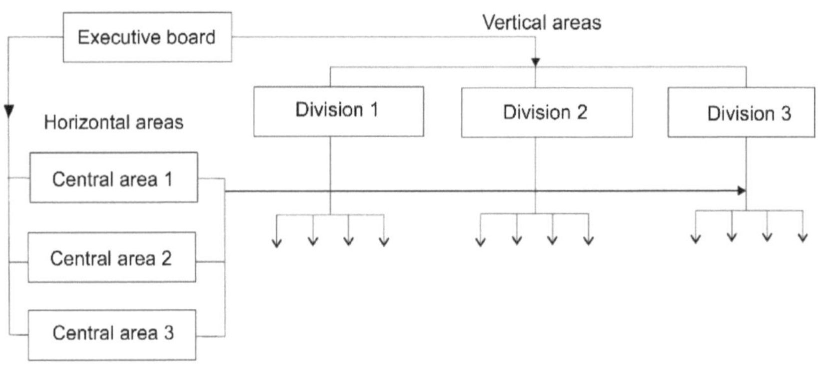

Figure 5.3: Matrix Organization

The project manager usually indicates the goal of the project. It determines what has to be done and when and forces the other bodies to decide who will do what and how. In doing so, he also forces the involved parties to think in an entrepreneurial way.

The matrix organization ensures flexible deployment of personnel. The project team members feel secure because they are not separated from their departments. However, the shared responsibilities may lead to conflicts between the departments and the project manager. If both have demands on an employee and cannot agree on who to follow first. This becomes worse the longer a project lasts and the more the employee identifies with the project.

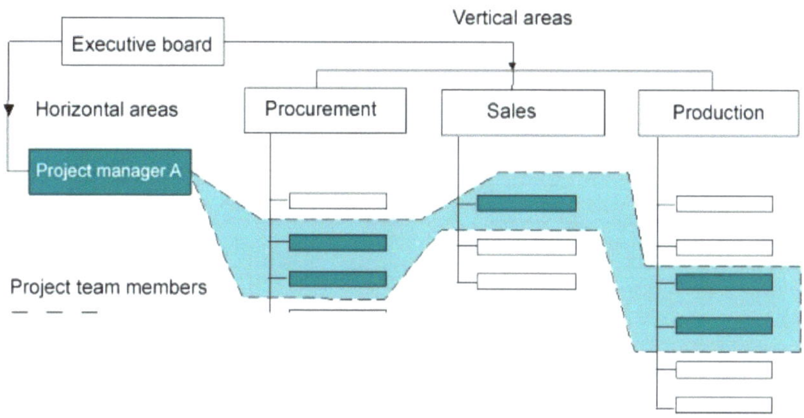

Figure 5.4: Project Matrix Organization

Much energy is required for these delimitations of competence. Therefore, through this division of responsibility and power, there is a constant potential for conflict between the line and project authorities. The greater the importance of the project in the company, the greater the problem of responsibility becomes.

5.2 Project Instances

The execution of a project also requires organizational and hierarchical structures in order to achieve the project goal.

In general, there are the following instances shown in figure 5.5:

- project organizator
- the steering committee
- the commercial processing (or controlling)
- the project manager
- the project group or the project team

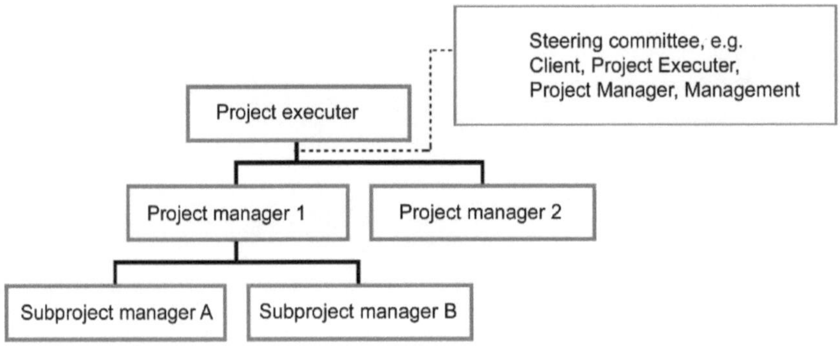

Figure 5.5: Project Organization with its Instances

Tripping Hazard

When team communication is bad!

Important information is not passed on to all team members intentionally or unintentionally, or only late.

- To avoid this, clear rules must be established by the project manager, for example how feedback is given. In project meetings, clear communication is ensured. In the case of conflicts, a solution is sought.

It is important to promote non-verbal communication alongside verbal communication (chapter 9.2).

Tripping Hazard

When team leadership is bad!

- The project vision is not formulated clearly enough.
- The team members do not have the necessary commitment to achieve the project results.
- The project vision must be reviewed.

5.2.1 Project Organizator

Project organizator means the responsible management of the planning, project planning, execution and control of one number of projects, i. e. always of several projects. This means that a project organzatorcan only be found in companies where several projects are processed simultaneously. There he has strong coordination functions.

The project orgaizator is always a member of the steering committee.

5.2.2 The Steering Committee

The steering committee is a liaison and arbitration committee made up of members of the executive board, the project organizator, the project manager and the client. The committee ensures that the implementation of a project is secured and that the concepts developed are accepted by the organisation. The steering committee is thus the principal of a project by approving individual results and representing the project group in dealings with the business community.

An important function is the arbitration function mentioned above. The steering committee shall deal with any disputes and contract disputes arising.

The steering committee appoints the project manager to his office and delegates the necessary competence and responsibility for the planning and implementation of the project to the project management.

5.2.3 The Project Manager and the Project Team

The Selection of the Project Manager

The compositionskills of the project manager has a great influence on the course of a project. However, very few companies have the possibility to select the most suitable candidate from a large number of suitable candidates. Most of the time, the most suitable person is already responsible for other projects or the ideal project manager' s/project team's egg-profile is not sufficiently fulfilled.

As a rule, you will always have to compromise between professional competence.

- Has he/she ever done such a project before?"
- Is he/she familiar with the technology and tools available?

and leadership skills

- Can he/she also guide the members of the project team through conflicts?
- Can he/she reach a sustainable consensus with the client?

have to agree on.

Figure 5.6 lists the most important core responsibilities of a project manager /5.1/ and the properties and skills required for this purpose.

PlanningTasks

- Clarification of the project objectives and participation in the development of the project definition,
- The drawing up of the project structure plan and structuring of the partial task and commissioning of the corresponding departments.

Manage the project

Plan and organize
- Targeted thinking
- Identifying and coordinating priorities

Controlling
- Results-related control of processes and activities

Stress tolerance
- High "frustration threshold

Manage customers

Target customer orientation
- Experience and feeling for the branch
- "Sale" of project results

Personality
- Convincing through one's own person (formation of trust)

Analytical capability
- abstraction capability

Concentration on the essentials
- certainty of judgement
- experience

Manage team

Leading
- Motivatuon of the team without management skills

Manage performance
- Support in case of technical problems
- Ability to delegate
- Motivation for performance
- Association of the various disciplines

sensitivity
- Effectively shape the strengths, interests and problems of team members
- Managing conflicts

Promoting communication

Figure 5.6: The Core Tasks of the Project Manager and the Required Competencies and Abilities

Tasks of the Project Manager

The project manager is responsible for the implementation of the project parameters defined in the project definition: time, costs and quality. He is also responsible for coordinating the project team.

The core tasks of a project manager therefore essentially consist of four main activities.

Coordination Tasks

- The release of funds within the project budget.
- The coordination of the project progress.
- The convening and management of project meetings.
- The application of the available resources.
- The design of the organizational structure and operational structure.
- Representation of the project internally and externally.

Inspection Tasks

- The planning and monitoring of project deadlines,
- Planning and monitoring of project cost development,
- Monitoring the progress of the project,
- To point out and counteract deviations from the plan,
- The examination, coordination and incorporation of necessary changes in the project plan.

Documentation Tasks

- Reporting on milestones in the steering committee,
- Ensuring the flow of information.

According to the project goal, therefore, are comprehensive and very different. They also depend on the size of the respective project. An important factor for the position of a project manager within a company is his or her involvement in the company hierarchy. The powers he has depend on this. Conflicts are inevitable due to unsettled competences.

The project manager must have excellent expertise in the most important project areas. In addition, they must also have knowledge of organisational methods and tools, planning and control, which are usually controlled by programs. In addition, there is commercial and legal knowledge.

Leadership skills are also very important. He must have great empathy, motivation, negotiation skills and persuasiveness. A high level of frustration is also important. In addition, analytical thinking is required, which includes an eye for the essential and a great ability to organise and improvise. Table 5.1 shows a job description for project managers.

Table 5.1: Job Description of a Project Manager

<table>
<tr><td colspan="3">Job Description</td><td></td><td></td><td></td></tr>
<tr><td colspan="4">Last Name, First Name, Date of Birth</td><td colspan="2">Function Designation</td></tr>
<tr><td colspan="2"></td><td></td><td></td><td>Range</td><td></td></tr>
<tr><td colspan="2">Company</td><td>Division</td><td>Group</td><td>Job No.</td><td>Tariff group</td></tr>
<tr><td colspan="3">Description of the Task:</td><td></td><td></td><td></td></tr>
<tr><td></td><td colspan="2"></td><td></td><td></td><td></td></tr>
<tr><td colspan="3">Employee reports to:</td><td></td><td></td><td></td></tr>
</table>

Continuation Table 5.1

<table>
<tr><td colspan="4">The holder of the position receives specialist instructions from:</td><td></td></tr>
<tr><td colspan="5">Directly subordinated employees (surname, first name, function, classification):</td></tr>
<tr><td></td><td></td><td></td><td></td><td></td></tr>
</table>

The holder of the position indicates his or her professional instructions (name, function):				
Holders are represented by:				
holder of a position represents:				
Authorities, power of attorney:				
The following tools belong to the task area				
(machinery, technical equipment):				
The following specialist knowledge is required to be able to perceive the task area:				

Table 5.2 shows the individual activities of a project manager and the average time required for them as a percentage within a project.

Table 5.2: Activities of a Project Manager with the Percentage Time Expenditure

	Activity Description	**Goal/Task Background**	**Share of time in %**
1	Development of a phase concept	Structured and clear project flow	20
2	Determination of the project team in coordination	Ensuring professional	10

	with their line managers	competence	
3	Creation of the work breakdown structure	Definition of work packages	20
4	Resource planning	Identifying resource bottlenecks, securing project progress	5
5	Initiation and monitoring of the necessary documentation	Information of the management	10
6	Organization and management of project meetings	Information of the project staff, planning of further activities	10
7	Preparation of project progress reports	Project controlling	5
8	Monitoring and coordination and securing of the technical know-how	Ensuring professional qualification, distribution of work and resources	5

Continuation Table 5.2

9	Initiation of timely invoicing	Liquidity monitoring	5
10	Preparation of a final report with post-calculation	Information of management, cost control, controlling	5
11	Project controlling in cooperation with the controller	Permanent project controlling with the aim of controlling and monitoring resources.	5

S	Special tasks	Special tasks, which belong to the activity of a project manager or result from the operational necessities.	∞

If a project is successfully completed, most of the time all involved parties are praised, if a project fails, the project manager will be the sole culprit. The success or failure of a project is in fact very much dependent on the person of the project manager, both professionally and personally.

Knowledge of suitable planning and control techniques is one of the basics for qualification as a project manager, but not the decisive one. His ability to cope with disruptions in the social psychological field is crucial.

At this point it is also pointed out once again that a project manager/employee always plays two different roles

- in the project he is contractor and/or client,
- in the line, he is subordinate/employee.

This can lead to conflicts.

5.2.4 The Composition of the Project Team and Team Building

The project manager is assisted by a team of specialists for the execution of the project.

It is often difficult to get the exact specialists you need. The departments try to retain their most capable employees, the employees themselves are often not happy to work on projects because the performance assessment and salary classification is generally carried out by the department head and not the project manager.

The question of an employee's ability to work in a team is usually the second most important question, but it is nevertheless important for the successful implementation of the project.

This is where the project manager's large management task lies. On the one hand, he involves the employees in the process and decisions very much, but he is also able to enforce the time and cost targets against the resistance of some team members.

The differences in opinion arising during the execution of the project are ultimately decided by him in order not to lose sight of the project goal.

The creativity of the team can be encouraged by several measures. So it is positive when

- Implement communication rules (for example, for meetings)
- an atmosphere of trust and collegiality,
- the "feeling of us" exists in the project,
- is communicated openly and comprehensively,
- Conflicts are detected and resolved as early as possible,
- good performances are recognized and passed on.

The project manager can only carry out the last point, however, if the company provides for it.

There is certainly a lot to be said about the way the project is handled. Unfortunately, the practical implementation depends on the people who are involved and there are always disruptive factors. However, some points should be taken seriously to prevent bullying in advance

- The team members show mutual respect and trust.
- Feelings such as anger are openly pronounced and not suppressed.
- Conflicts and problems are openly addressed and solved.

- Active listening is practiced.
- All are equal partners and no one is doomed.
- Only constructive criticism is made. It is solely for the purpose of progress and cannot be taken personally.
- Different opinions are perceived as a contribution to the solution of the problem.
- Consensus is sought in the team. All team members abide by decisions once made.
- All follow the "rules of the game": good preparation before meetings, completion of the given tasks on time, punctuality.
- Information about the project will be communicated to everyone.
- The activities of all team members are known to everyone else. No team member performs activities without prior consultation with the project manager.
- The tasks such as logging, monitoring lists, etc. are distributed fairly.

This is a wish list. Nevertheless, it is important to try to practise such behavioural patterns within the team in order to make the project successful. Further behaviours and solutions to problems in the team are described in chapter 9.2 under Business Coaching.

5.3 Test Your Progress

Now you can check your knowledge again with the help of table 5.3.

Table 5.3: Chapter 5 test. Which statements are correct or false?

		correc	**false**

		t	
1	In principle, projects are initiated by supervisors.	☐	☐
2	Anyone can be a project manager.	☐	☐
3	The project manager is responsible for the implementation of the project.	☐	☐
4	The initiator of a project usually has clear ideas about all important aspects of the project.	☐	☐
5	The project team must clearly define the project before starting.	☐	☐
6	Project management is hardly different from other management applications.	☐	☐

Continuation Table 5.3

7	The time constraints of projects pose particular challenges to the project manager.	☐	☐

1	**2**	**3**	**4**	**5**	**6**	**7**
f	r	f	f	r	f	r

Tasks

- Distinguish the three project organisation forms.
- List four instances that are usually involved in the execution of projects.
- Explain what a steering committee means and what kind of task it has.

- What professional and personal requirements must a project manager have?

Here you can book a free Strategy Session or write me if you like this book and have suggestions or questions.

Click here for free bonus material on the book.

Also visit my blog „Selbstführung & Produktivität in German“. I'll help you get better results.

6 Project Execution

6.1 Activities in the Execution Phase

During the project implementation phase, the project manager coordinates all individual steps of the project. This includes both the areas of responsibility mentioned above in the presentation of the project manager's position and the special leadership taskn in the project's day-to-day work.

- Control of work processes to ensure their scheduled completion,
- provide feedback to the employees,
- Conduct negotiations to provide materials, equipment and services,
- solving conflicts within the project team and negotiations between the project team and outsiders.

These key tasks require different skills of the project manager. In this chapter and in chapter 9 some techniques and tools are presented, which help the project leader to carry out the project.

Control is the most important activity during the execution of the project. The most important tool for this control process is the project plan, for example as a Gantt chart. It develops the three project parameters, the individual steps with specifications, the schedule and the budget. These are the yardsticks by which the progress of the project and thus the performance is measured.

6.2 Control of Work Processes

Tripping Hazard

If the progress is monitored too inaccurately and/or too late!

- You do not enter personnel expenses and expenditures or start and end dates for certain activities.
- You do not pass this information on to the team members in time.
- Performance can only be monitored by setting standards. The responsibility for monitoring must be laid down in writing.
- The control of the work processes consists of three steps:
- Set standards,
- Monitor performance,
- Take corrective action.

6.2.1 Set Standards

The standards for the project are set in the detailed project specifications, which are created in the planning phase. The project manager keeps an eye on these specifications at all times and ensures that employees do the same.

If the project deviates from the original specification, the predicted success of the feasibility study is no longer guaranteed. The probability that the project or the project result could not meet the performance criteria is very high.

In order to determine whether the quality, time and cost standards set out in the specifications are actually met, and also to facilitate overall control of the project's progress, the project manager has a range of tools at his disposal. The relationship between the schedule and the project's own progress can be tracked using the GANTT charts and PERT diagrams created during the planning phase.

In the following, four further presentation options are described which are useful for project control:

- Control Point Recognition Charts
- Project control charts
- Milestone charts
- Budgetary control charts

Control Point Recognition Charts

A helpful technique for project control is the targeted notation of possible error possibilities, which could occur with the three project parameters. The project manager and the project team decide when and how they will learn that something is not going as planned. Then they can make a plan of how to correct the error when it actually occurs. This prepares them for unpleasant surprises and allows them to react more quickly to problems. A control point detection chart summarizes this information in a clear and concise manner (table 6.1). The listing is already started in the offer phase and is continuously supplemented during the course of the project.

Another useful tool is the project control chart, which provides a quick overview of budget and schedule deviations. The planned dates and costs are compared with the actual ones. The deviations for each individual step and for the entire project are thus clear.

To prepare a project control chart, all the individual steps on the GANTT chart are listed. Then you use the schedule and budget by assigning the planned duration and expected costs to each individual step.

Table 6.1: Example of a Control Point Recognition Chart

Control Element	What could not work?	When and how will I find out?	What am I gonna do about it?
Quality	Workmanship is defective.	After personal inspection of the work of a completed production step.	Work must be repeated.
Costs	Costs of a single step exceed the budget planned for this purpose.	As soon as purchase contracts have been negotiated.	Search for a replacement supplier first, then consider replacement material.
Time	Duration of a single step exceeds the planning duration.	By closely monitoring the progress of work along the critical path.	Look for ways to improve performance, try to save time in later steps, if the budget allows you to allow overtime hours.

After a single step has been completed, the actual data is entered. The variances are calculated and added (table 6.2). The single-step durations with * are not on the critical path and could therefore be taken from the total.

This technique can be implemented on any PC in MS Project® or simply as a list form. You can then use it to generate the progress reports.

Table 6.2: Case Study Hall Construction, Project Control Chart in T€; Single Step Periods with * are not on the Critical Path

Subunit	Costs				Time			
	budgeted costs	actual costs	variance	total	duration planned	duration actual duration	variance	total t
1. Complete work plans	10	9	-1	-1	15	15	0	0
2. Obtain building permit	1	1	0		16	15	-1	-1
3. Concrete foundation	50	55	5	4	5	3	-2	-3
4. Erect walls and roof frames	44	42	-2	2	5*	5	0	-3
5. Finish the roof	25	3 0	5	7	5*	6	1	-2
6. Install window	15	15	0	7	1*	1	0	-2
7. Set up outer sides	76	67	-9	-2	10	9	-1	-3

Continuation Table 6.2

8. Paint outer	6				3			

sides								
9. Lay the electrical net-work	13				10*			
10. Install heating/air conditioning system	36				5			
11. Finish in-sulation	15				5			
12. Install in-ner lining	18				5			
13. Install inte-rior doors	14				5			
14. Painting of interiors	7				3			
15. Install electrical equipment	3				2			
16. Perform cleaning	3				3			
17. Laying floors	8				2			

Milestone Chart

A milestone chart gives a rough picture of the project's schedule and control data. It shows the key events (miles or milestones) that are clearly identifiable to outsiders or require approval before the project can continue. In MS Project®, a milestone chart can be created using a simple filter function.

In general, there are few milestones in the project. Because of the overview character, a milestone chart is not very helpful during the planning phase, when more information is needed. However, when the project is completed, it is very useful because it provides a concise summary of the project's progress (table 6.3).

Budget Control Chart

There are two variants of the budget control charts. The first is very similar to the project control chart. It compares the budget with the actual costs of the individual steps of the project. The other variant is a diagram that compares the planned costs with the actual costs incurred. Either bars or lines are used for this.

The bar charts compare the planned costs with the actual costs incurred for each individual step, while a line chart cumulatively compiles the planned and actual costs for the entire project.

Small deviations, as seen in figure 6.1 and figure 6.2, are perfectly normal. Only in the event of large deviations will the experienced project manager intervene and take action.

Table 6.3: Case Study Hall Construction, Milestone Chart

Milestone	Planned Completion	Actual Completion
1. Foundation completed	5. April	2. April
2. Frame completed	10. April	7. April
3. Exterior sides finished	25. April	
4. Electrical network is installed	31. April	
5. Heating and air conditioning installed	4. May	
6. Interior finished	22. May	

Small deviations, as seen in figure 6.1 and figure 6.2, are perfectly normal. Only in case of large deviations will the experienced project manager intervene and take measures.

6.2.2 Monitor Performance

The heart of the control process lies in the monitoring of the running operations. The project manager learns what has actually been done and how the project progress is in line with the plan. This tells you whether a correction is necessary. Common monitoring measures are

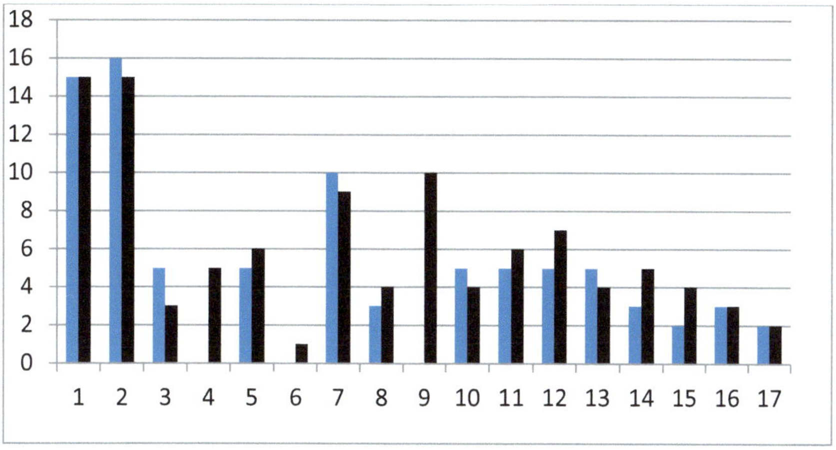

Figure 6.1: Case study: Hall Construction Project; Budget Control Chart; Black: Budgeted, Blue: Actual Costs

- Inspections,
- Progress meetings,
- Quality tests and
- Examinations.

More than one source of information is needed for effective control. In addition to written records, the project manager was kept up to date with a mix of inspections, progress reviews, quality tests and audits.

6.2.2.1 Inspection

Inspection is the most common monitoring measure for projects. They are carried out by inspectors and the project manager. Work processes are monitored as they are carried out on site or in the production facility. Inspection is a very effective way to learn,

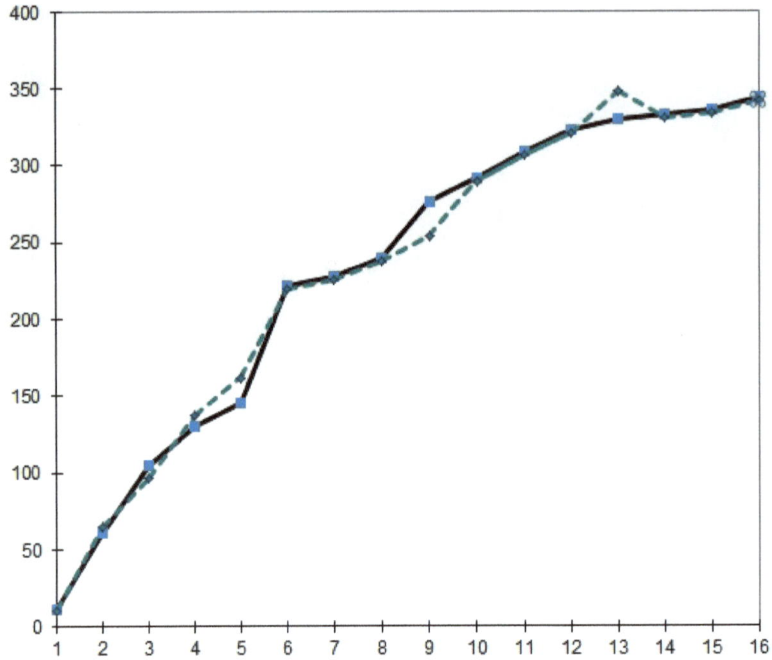

Figure 6.2: Case Study: Hall Construction Project; Budget Control Chart; black: Budgeted, dshed: Actual Costs

- whether the project specifications are adhered to,
- whether unnecessary waste takes place or
- whether dangerous working conditions prevail.

Inspections are generally unannounced and unpredictable, but can also be open and direct. Questions can be put to the implementing employees, which often enable the project manager to draw important conclusions.

6.2.2.2 The Progress Review

Progress reviews are discussions between the project manager and the employees responsible for each individual step. They can be guided with groups or individuals, face to face or by telephone. They can also be submitted in writing to external projects. Progress meetings are usually held at regular intervals, for example daily, weekly, monthly.

They can also be called if a performance sample is noticed or if an important step in the project begins or has been completed. Meetings called by the project manager or the employee responsible for the work supplement the planned progress meetings.

Guidelines for Progress Reviews

Normally, three issues are on the agenda in the progress meetings:

- the comparison of planned and actual progress,
- review of problems that have arisen and how to overcome them,
- the discussion about upcoming problems and their possible solutions.

The different roles of the project leader during a progress meeting.

The task of the project leader is active during a progress review to find out the status of the project and to influence the further progress, even if this is necessary. During the meeting, the project manager plays very different roles (figure 6.3).

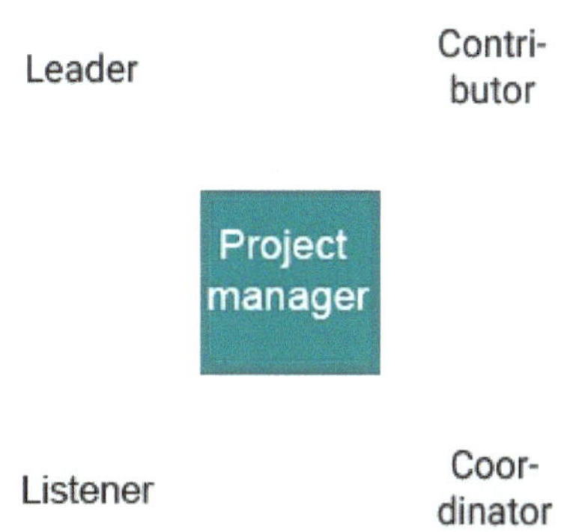

Figure 6.3: The Roles of the Project Leader During the Progress Review

The Listener

The project manager should listen very carefully when informed about progress, deviations from the plan, problems and proposed solutions.

It is not only important what is said, it is also important how something is said! Is the employee excited, frustrated, discouraged? The answers can be explained by questions. Recovering what has been said can help the listener to gain more precise impressions and feelings on the one hand, but also to clarify the actual problem (chapter 9).

The Contributor

Progress is progressing according to plan at many interim meetings. But there will also be problems to overcome. If this is necessary, the project manager should actively contribute to the solution, for example by pointing out further possible solutions to the project team. The experience and knowledge of the project manager is necessary to advance the project.

The Coordinator

An important role of the project manager is that of the coordinator: the individual steps of the project are combined to a functioning whole. What is neglected? Is any work done twice? How to deploy personnel who become vacant?

The Executive Manager

The most important role of the project leader is that of the leader. He must always keep the team's efforts focused on achieving the project goal: the completion of the project within the budget and time frame, as well as adherence to the specifications. Good achievements have to be acknowledged, weak achievements have to be corrected, interest and motivation in the project have to be kept high.

After all, there is a unique dynamic in each group and team. The group characters shown in figure 6.4 appear in each group.

If one thinks by a lucky coincidence to have gotten rid of an unpleasant charakter, another group member emerges, who takes over his role.

Figure 6.4: The Group Characters /Source: Unknown/

Table 6.4: The Possibilities to Use the Different Characters in a Positive Way for the Team

1	**The belligerent (the grouch)**	Stay objective and calm. The group cause his allegations. to be refuted.
2	**The Positive** **(Pushy One)**	Summarize the results.
3	**The Knower of All (Insider)**	Invite the group to comment on its allegations.

Continuation Table 6.4

4	**The Frequent Speaker**	Interrupt tactfully, set speaking time.
5	**The Shy One**	Ask easy and direct questions. Strengthen his self-confidence.

6	**The Denier**	Recognizing his knowledge and experience.
7	**The Thick Skinned (Uninterested Parties)**	Ask for his work. Give examples from his field of interest.
8	**The Sublime (The "big" animal)**	Do not criticize directly. Use „yes, but"- technology.
9	**The Interrogator (Have a Point)**	Return his questions to the group.
10	**The Team Spirit**	Trying to wake him up.

6.2.2.3 The Quality Test

Quality tests are usually defined in the project specifications for quality assurance. Typical tests include, for example, loading tests on mechanical components and materials.

6.2.2.4 The Examination

The audits oe examinations can be carried out during and after a project has been completed. Usual areas for audits are accounting, purchasing and security practices, maintenance procedures or disbursement powers. The auditors must be experts in their respective fields and come from outside, i. e. they must not be members of the project team. After the thorough audit, an audit report or examination report is drawn up which describes in detail the results and shows everything that deviates from established guidelines, approved practices or reasonable business practices.

6.2.3 Take Corrective Action

During the progress and performance monitoring, there will always be times when the project will fall short of plan. This generally always requires corrective action, but the project manager should also be careful not to intervene too quickly. Some defects will correct themselves. Sometimes the project will lag behind the plan, sometimes it will progress faster than expected, but it will eventually be "on track". It would be unrealistic to expect steady progress on schedule.

If the quality does not meet the specifications, the work is usually repeated or improved. But even in these cases, however, decisions should not be taken too hastily. Perhaps the work will meet the requirements if it is examined in more detail. It should be carefully checked to see if the actual deviation is correct and whether it could have a serious impact on the project. If something is then repeated or reworked, it is a single decision and not an automatic process.

When the project actually starts following the plan, there are several ways to deal with it.

First of all, it is checked whether the remaining work can be done faster and the time can be obtained.

If this is no longer possible, employees can be offered additional incentives to do the rest of the work even faster. Such allowances can be justified by a direct comparison with the potential loss in the event of late settlement.

If this does not seem possible or effective, the increase in resources can increase work performance, for example through overtime or the employment of additional workers. This may cost more, but it can prevent losses or penalties due to missed deadlines. Table 6.5 lists some of these measures.

Table 6.5: Catalog of Procedures for Corrective Actions

1	**New negotiation**	Discussing with the customer a possibility of extending a deadline or increasing the budget.
2	**Catching-up**	If time is lost with connected individual steps, the budget and schedule for the remaining work must be checked. Perhaps savings can be made there in order to meet the cost and time targets.
3	**Limitation of project scope**	Unimportant project steps (wish list) can be deleted to reduce costs and save time.
4	**Increase resources**	More employees and machines can be used for the project to keep an important deadline. The costs for this have to be compared with the importance or the benefit.
5	**Search for a replacement**	If something is not available or more expensive than expected, replacing it with comparable materials or equipment can save money or make work possible in the first place.
6	**Alternative sources**	If a supplier is unable to deliver within the specified time or cost limits, a replacement for its scope of delivery should be sought immediately. However, the alternative offered by the supplier can also be accepted before it is replaced.

Continuation Table 6.5

7	**Part deliveries**	Sometimes a supplier with partial shipments can help to keep a project on schedule and deliver the rest of the delivery later.

8	**Bonuses**	Going beyond the limits of the original contract or offering a supplement or similar incentives can still make it possible to deliver on time.
9	**Demand compliance with the contract**	Sometimes it is possible to achieve the desired result by insisting on the demand, also through the threat of a penalty, to adhere to agreements made. The best option is if this is already part of the contract. It may be necessary to ask the management for support.

When the project starts to exceed the budget, it is also necessary to consider how to make savings in the work still to be done. Otherwise, the scope of the project must be limited or more funds must be negotiated with the client.

6.3 Managing Feedback

A project manager has many opportunities to provide feedback to employees, i. e. feedback on their work or behaviour. Through feedback, the employee learns how his or her behaviour affects others and the success of the project. This can promote good performance or improve poor performance. Figure 6.5 shows a feedback sheet that leads to an improvement.

When giving feedback, it is very important to limit oneself to actual actions and concrete results, because one can only guess the intentions of others.

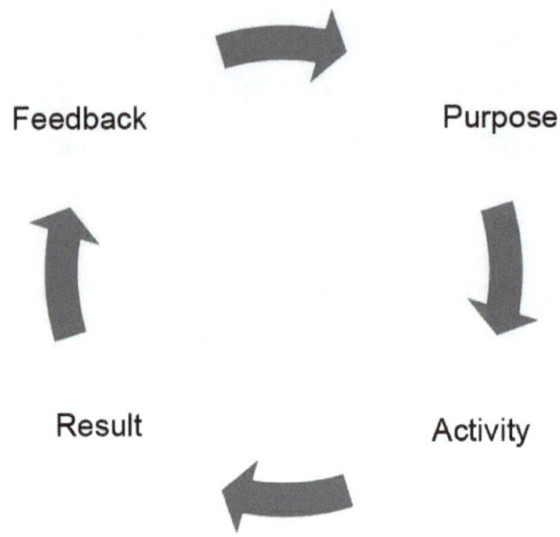

Figure 6.5: The Feedback Arch

If you give a positive feedback, you describe the actions and results directly and concretely and endorse a positive evaluation. For example, it is said,"Thanks to your efforts last night, the project step could be completed in time. I appreciate your extra effort."

In the same way, negative feedback should be formulated. But another element should also be added to help employees deal with similar situations more appropriately in the future.

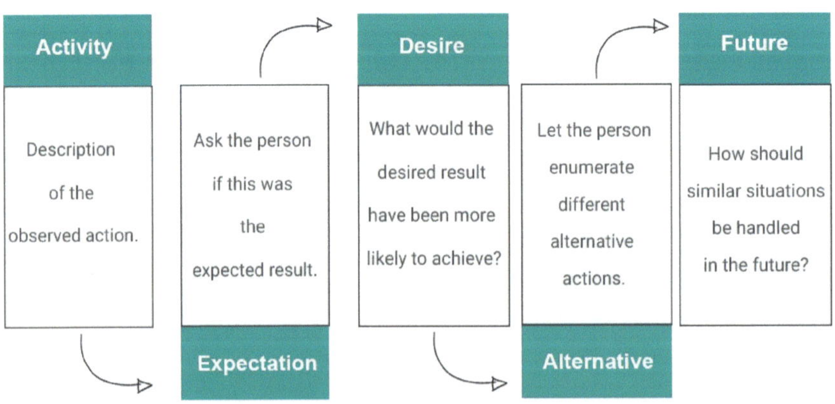

Figure 6. 6 Dealing with Negative Feedback

However, this procedure only makes sense if the employee has unintentionally acted inappropriately. If there is a defiant “yes” answer, it will certainly not be possible to proceed in this way. Then other methods must be used to change the behaviour in terms of the project goal (Chapter 9).

6.4 Project Communication

6.4.1 Conduct Project Meetings

Project meetings are a very important part of the project manager's work. They take up to 20% of the time spent on each project.

Negotiations are also an important tool to achieve the success of a project, for example by identifying problems at an early stage.

In meetings, it is very important to work together if you want to achieve a positive goal. It must be ensured that the parties are equal and that everyone is aware of this. So no party has the power to force a result. That's what makes up the negotiation skills.

6.4.1.1 Communication Rules for Meetings

The Preparation

The meeting leader must know the purpose of a meeting or negotiation in detail and must also communicate it to the employees as accurately as possible.

An important part of the goal definition is to consider what to do if no agreement can be reached according to the project leader's ideas or project requirements. The influence on the outcome of such a negotiation is based on attractive alternatives which can then be presented. The easier it is for the project manager to abort or postpone negotiations, the stronger his negotiating position becomes.

All participants of a meeting, including the project manager, are obliged to take care of the "exercises", i. e. they must avoid appearing unprepared in the meeting. No employee should come unprepared. All participants should take the time they need to prepare, even if they have to ask for a stay.

Minimization of Perception Differences

The picture of the event that a person makes is based on his or her "history" and experience. This may differ significantly from that of the negotiating partner. That is why one should never assume to know the other's point of view. It is only through scrutinizing that ambiguities can be removed and agreement reached. Therefore, it is very important to define a thing very clearly, so that the opponent can confirm or correct the picture.

This is aided by the questioning techniques presented in chapter 9. The situation is precisely grasped by these questions. By careful observation you can see what is going on in the opposite direction and then take the appropriate measures.

Listening

Active, attentive listening is an obligation for effective negotiation. The chairperson must let the others have their say. When he talks about 50% of the time, he clearly doesn't listen enough.

This also includes respecting quiet breaks. The new impressions must first be processed before a continuation makes sense. No one should be tempted to fill these creative breaks with speeches.

Taking Notes

It is necessary to note down what is discussed and what is decided. In the case of high demands, it makes no sense to rely on memory alone. For this reason, the agreements reached are summarised in a memorandum. This memorandum or protocol must specify not only the measures but also the persons responsible and the deadlines. If the project manager cannot do it himself, he or she decides to have someone do it before the meeting.

Contributing Creativity

Too early termination of the meeting or nagging criticism of the employees' statements dampens the spontaneity of those involved. A little time should be allowed for alternative or unusual solutions from the outset. During such a discussion, all ideas can be presented in a value-free manner, i. e. without criticism. All negotiations, but also all further cooperation can benefit from such a creative approach.

Supporting the Other Party

Good negotiating partners recognize that the problem of the opposing party is also their own problem. To this end, one moves into the other party and works out a solution that is acceptable to everyone. After all, a deal only lasts if it's carried by everyone.

Making Compromises

"Give something for nothing" should be avoided. At least a promise of goodwill or a future repayment must remain for giving. Even if the promise you have made is not worth as much as the other person, there is always something that is more important to you than the other side.

Bringing in Apologies Quickly

If you have offended yourself in the choice of words during the trial, an apology is the quickest and safest way to get rid of negative feelings in other people.

This is not only necessary when it comes to personal forgiveness. An apology for the current situation can also be effective.

So hostile remarks should not contribute to a bad atmosphere of discussion. Hostility directs the discussion from the essentials to a level of self-defence at which one wants to harm the other.

Avoiding Ultimata

An ultimatum always demands that the opponent either surrenders or fights to the bitter end. Neither one nor the other outcome will be conducive to a future positive cooperation in a project. Therefore, you should avoid cornering someone. This happens, for example, if you only offer the other party alternatives that are unacceptable to the other party. A back door should always be left open to the other party.

Setting Realistic Deadlines

Many meetings take too long because there is no deadline pressure (timetable for the meeting). A deadline requires both sides to use their time effectively. A realistic deadline pressure encourages both sides to make concessions and compromises.

However, deadlines must be avoided which are not realistic and cannot be met.

6.4.1.2 Organization of Meetings

The more carefully meetings are prepared, the more effective they will be for all parties involved. The project manager will announce the topics of the meeting as early as possible so that all team members can prepare themselves accordingly and make any wishes or corrections in advance. Table 6.6 shows the example of an invitation to a project meeting.

For example, if the next meeting is monday at 2 p. m., a realistic deadline for the receipt of all applications and proposals for topics is 5 p. m. By thursday at the latest, the proposals submitted later will be considered in the following discussion rather than in this discussion.

This does not apply to highly explosive, sudden problems in the project. A meeting can be thrown over the top of the board for that. But that should definitely be the exception.

If everything goes according to plan, the agenda for a meeting on Monday can be announced on friday.

The basic rule for the number of agenda items is: 5±2 is a reasonable quantity. It is important that there is no agenda item without action. This belongs in the category "Miscellaneous".

The duration of an efficient meeting should be about 1.5 hours.

Once an agenda item has been processed, it is summarized again by the project manager or an employee. The responsibility is distributed exactly (who does what?) and the completion date is given (until when?). In order to be able to follow this action, the project manager will make a note "follow-up". The person responsible will inform us of the execution or completion of the item, if necessary also in writing.

The protocol, which can be kept by alternate team members, contains the most important, minimum information of the items on the agenda, which results the discussions yielded and which actions were determined. This may also be necessary literally. For this purpose, the person can ask for the wording to be dictated literally.

If a meeting is to be regarded as an "open-ended event" due to the upcoming problems in advance, all team members should also be aware of this beforehand in order to take appropriate precautions, for example to cancel private appointments.

The time assumptions are necessary for scheduling the meeting. This makes it easy to identify and schedule controversial or detailed discussions. Whether the times should then be announced in the invitation must be decided individually. When they are announced, they are visible to everyone and can be used as a control element by the meeting leader.

6.4.1.3 Presentation Techniques

Presentations today are subject to very high quality requirements due to the many possibilities of multimedia. For example, animations can be integrated into a presentation with very simple means, which are projected onto the wall with the help of a projector.

This is always necessary in a running project. Some basic rules should be observed:

- Do not overload your slides with content. 5±2 points the listener can accept well, with more he feels overwhelmed.
- Write large enough so that those in the last row can still read your slide. Printed pages cannot simply be copied as a folie. The result is usually insufficient.

Table 6.6: Example of an Invitation to a Project Meetingg

Start:	14:00	End:	16:15
Partici-pants:	Project manager, Mr. X, Mrs. Y		
Uhrzeit	What?	wer?	Zeit
14:00-14:05	Welcoming speech	Project manager	5 min
14:05-14:30	1. Comparison of planned and actual progress	Project manager	25 min
14:30-14:40	Discussion	all	10 min
14:40-	2. Looking back on	all	35 min

15:15	problems encountered and how they were overcome		
15:15-15:20	Discussion	all	5 min
15:20-15:45	3. Discussion about up-coming problems and proposed solutions	all	25 min
15:45-16:00	Discussion	all	15 min

Continuation Table 6.6

16:00-16:15	4. Miscellaneous	all	15 min (max.30 min)
16:15	Summary and goodbye	Project manager	

- When your listeners hear the presented topic for the first time, give them a chance to enter the ma-terie. This requires a clear definition of what is to be learned during the presentation and how you will proceed.
- Find out in advance whether there is time for questions and discussions after the presentation. If this is not the case, you must take it into account in your lecture planning.
- Do not overload your presentation with things that only affect the topic. Interested listeners will contact you to find out more details about you that are of particular interest to them.

Table 6.7 shows a general presentation structure that can be used to prepare presentations very successfully.

Like a meeting, a presentation in itself is a project and needs to be structured in a meaningful way. The effective presentation requires four steps: planning, preparation, training and presentation.

Table 6.7: General Presentation Structure

Plan-ning	Inform yourself beforehand about your listeners by asking questions about their level of knowledge, experiences, needs and goals.

Continuation Table 6.7

	Define the lecture purpose and the lecture objective. Do you just want to inform or convince, to motivate to act, to sell something, to teach someone or educate?
Prepa-ration	Create a positive mood for yourself by giving your statement the correct value, imagining your success, imagining the re-action of your listeners and motivating yourself.
	Prepare an introduction that attracts attention by asking a question that speaks directly to your audience and involves them in the topic, telling them about a relevant event that is as up to date as possible.
	Illustrate and support your arguments with explanations and illustrative material such as statistics or analogies, experi-ments or other demonstrations, events, existing results and evidence.
	Prepare a catchy closing statement by challenging the audi-ence to do something, maybe even provoking something, or by making a motivating statement, repeating the main bene-

	fit of your presentation in a compelling summary.
Trai-ning	Strengthen your self-confidence by keeping in mind that through your competence you have the right to present this presentation and to make this interesting topic accessible to your listeners.
.	Train your presentation and check your visual material for clarity and memorability, meaning, visibility and visual im-pact, quality.

Continuation Table 5.2

	If you are untrained or "something is at stake", practice your presentation in front of trusted listeners. Ask for feedback and advice on stylistic and content issues such as your presentation style, your posture and body movements, your language (say "uh"again and again), an effective introduc-tion and a memorable ending, clear arguments (5±2) and credible justifications, etc., logical thinking ("red thread") and the presentation of the achieved results.
Pre-senting	Also during the presentation, make it clear that you have been selected, are responsible and take the opportunity to present your results here.
	You will be confident and self-confident in that you will be able to convey an important message to the audience by mastering the topic, having prepared and practiced your presentation well.
	The first impression is the most important, so make eye con-tact with the listeners, pay attention to a self-confident body language, relax, take care of a well-groomed appearance.
	Get in touch with the audience by being correct and, of course, by saying "we" and "us", not "I"when you are work-

	ing in a group, or even "them", by addressing your listeners directly again and again to arouse your interest.

Continuation Table 5.2

	Attract the viewer's attention by showing enthusiasm, using all five senses to describe the subject, for example, using vivid words next to the slides or including the imagination of your listeners (chapter 9), for example, telling a story that decorates the subject, expressing yourself clearly and concisely and speaking with enthusiasm, and possibly paying attention to the corrective body movement.
	To confirm and improve your presentation style, you measure the success of your presentation against the feedback of your audience, identify the strengths and the parts of the presentation that can be improved. Note these improvements for your next lectures.

6.4.2 Successful Telephoning

A lot of telephone calls have to be made during the project work. Those who observe the basic rules of communication on the handset can effectively present themselves and their concerns.

Discussion Preparation

A good preparation and a clear goal orientation are guaranteed by the interlocutor, because it signals professionalism and attention. This can be helped by a standardised list of options, which is tailored to your personal requirements:

- Which questionn do I have?
- Which questions can I ask?
- What is the latest correspondence?

Every accurate answer makes you more confident.

The First Few Seconds

The first few seconds of a phone call decide on sympathy or antipathy, especially if the person you are talking to is not yet known. The greeting is therefore of crucial importance.

The message, when you are called or called, starts with a greeting, for example "Good morning","Good day" or "Good evening", then the naming of your own name, possibly with first and last name and the name of the company, possibly also the naming of the project or department. And all this is being said slowly and clearly.

It's not professional to deal with "Hello?" to report. The caller wants to know first of all whether he is connected correctly, especially when he calls for the first time.

It is just as bad to chant the names and designations like an annoying duty or to pronounce them in the chord.

Active Listening

It's best to be brief. After a maximum of three to five sentences an anchor question "What do you think of it? That's what the other person has to say. In this way, important information about the needs and mood of the partner can be obtained. This makes it easier to assess how to proceed. This is particularly useful if the person you are talking to is in a meeting or is travelling by car.

The Posture

Upright sitting or even standing makes the voice freer and firmer because you can breathe better. For this purpose, the handset should not be jammed between the shoulder and cheek, even if you have to write something down during a conversation, as this makes the voice appear more involuntarily pressed. Professionals use headphones.

Recording to an answering machine in particular should be calm, clear and friendly. In addition to the request, it is important to know the time of day and your own telephone number, which should be spoken for a long time and repeated once.

The Discussion Atmosphere

A pleasant atmosphere creates a pleasant atmosphere. The best way to do this is by naming the name of the conversation partner. This creates a familiar situation.

Also to spread words of personal appreciation, for example: "Thank you for calling back immediately.", create a good atmosphere.

But all exaggerations should of course be avoided. This would then appear to be a calculation and thus false.

If one of the interlocutors gets involved in a complaint, for example, his own friendliness can nip most of the attachments in the bud. However, such situations require a high degree of empathy. It is precisely these difficult moments in which an employee can prove how customer-friendly and flexible he is.

Conciseness

Accuracy is a crucial criterion for the success or failure of a telephone call. Misunderstandings can have fatal consequences, especially if agreements are made on project agreements. That's why we need to listen very concentrated.

If you have the feeling that both callers say the same thing, but do not mean the same thing, you should ask politely and clarify the arrangement until both callers know exactly what they expect from each other.

Address All Senses

If a situation or a product that the other party does not know has to be described, it should be described as clearly as possible.

You should try to address all five senses, the feeling, the sense of hearing, hearing, feeling, smelling and tasting. Although the telephone is not able to transmit a smell, a taste and usually not an absenteeism, the verbal description alone can create its own character.

Short Summary

The conclusion of a telephone call always requires a short summary. It's also an elegant way to signal the other person that the call is being ended, such as the question,"Is there anything else I can do for you?".

If there are no more questions left, the phone call is summarized in clear and concise words. It should then also be clarified whether it helps both if the result is recorded in writing.

Informedness

If the telephone of a colleague or even the project manager needs to be used, each team member in a well organized project should always know where the desired caller is staying for how long. Disinterested demonstrating sentences like "He's not there yet" or "I don't know when it's coming" should never fall.

Moods

Your own bad mood or a bad day is basically your own problem, never that of the person you are talking to. Employees who are unable to control themselves and bring in their private anger cause irreparable damage.

Table 6.8: International Alphabet to Spell;

A = Alpha	B = Bravo	C = Charlie	D = Delta	E = Echo	F = Foxtrott
G = Golf	H = Hotel	I = India	J = Ju-	K = Kilo I	L = Lima

			liette		
M = Mike	N = No-vember	O = Oscar	P = Papa	Q = Quebec	R = Romeo
S = Sierra	T= Tango	U = Uni-form	V = Victor	W = Whiskey	X = X-Ray
Y = Yan-kee	Z = Zulu				

Telephoning Times

There are times that are not suitable for the call. For example, if the person you are talking to is just before a conference, or shortly before closing time or on the last day before your planned holiday.

Also, a one-time permission to call in the evening privately or on a mobile phone is not a carte blanche to do so in the future.

The International Alphabet

Understanding is best when spelled according to the official norm. This helps the conversation partner .

It's important to note things down correctly and not have to be queued up all the time. Table 6.8 shows the official international alphabet.

6.5 Time Planning and Priority Management

The gods gave us a long life,
but we shortened it.
Seneca

Time planning is the core of any working method and successful self-management.

Often enough we hear during project meetings or at the distribution of the upcoming tasks that the employee is "already up to the top". One member of a project team commented with the example of his whiskey bottles,"If I have a shelf for my ten favourite bottles of whiskey and I put another eleventh on it, one falls off."

That's the truth. But if it were our favourite whiskey bottles, we would solve the problem in a different way. The following answers are spontaneous in table 6.9.

Table 6.9: Problemlösungen bei zu vielen Taskn

On the Whiskey Shelf	In the Project
I drink up a bottle that has been started and can replace it with the new one.	I finish a task and have time for the new one.
I'm giving away one of the bottles, maybe those I don't like so much anyway.	I reexamine the priorities of the tasks at hand and delegate the less important tasks.

No one would drop the bottle. In projects, however, this is done by putting their heads in the sand and forgetting many of them.

In many projects, the employees actually have enough to do. Nevertheless, there are always opportunities to retrain oneself and to reorganize one's own time and priority management.

Again, define your goals!

In chapter 9, the definition of goals is dealt with even more intensively. Here, the most important points are dealt with on the basis of Covey, Merill, Merill /6.2/, /6.3/ time quadrants according to the EISENHOWER principle (figure 6.7), which summarize all that has been taught in time planning and priority management for decades and summarizes it in a nutshell.

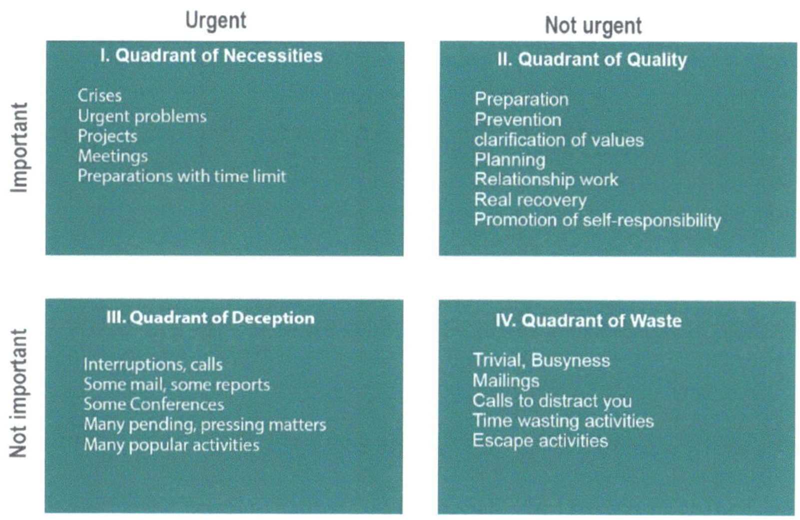

Figure 6.7: Time Quadrants According to Covey, Merill, Merill /6.2/, /6.3/.

In order to get an overview, we divide all activities, professional or private, into four quadrants.

I Quadrant of Necessity

The activities in this quadrant are both urgent and important. Here we talk to the customer, keep an appointment, repair a defective machine, get heart surgery or help a crying child who has been injured.

If we ignore him, we'll be buried under a mountain of duty. But we also have to realize that many tasks are only urgent due to hesitation or lack of planning and anticipation.

We cannot wait and see if an aggregate fails or if we begin to comfort a child the next day. This must be done immediately.

II Quadrant of Quality

Quadrant II receives activities that are not urgent but important.

This is where we make our long-term plans, prevent problems to be overcome, demand responsibility from ourselves and others, broaden our horizons and improve our skills through reading and continuous professional development, prepare ourselves for important meetings and presentations and engage in deep relationships by listening to them correctly.

The more time we spend in this quadrant, the greater our ability to act becomes. Ignoring this quadrant causes the quadrant I to swell and leads to stress, exhaustion and deep life crises.

Conversely, great commitment in Quadrant II shrinks quadrant I. Planning, preparation and prevention ensure that many things do not become urgent.

Quadrant II doesn't affect us, we need to affect it.

This also applies to private life. How often do we think we can relax with a computer game, or let the TV drizzle over us instead of taking a walk or going to bed.

III Quadrant of Deception

Quadrant II is almost a copy of Quadrant I and contains things that are urgent but not important.

The hustle and bustle of the urgent creates an illusion of importance. But the corresponding activities are at most important for someone else, many calls, meetings and random visitors fall into this category.

We spend a lot of time in Quadrant III and, in the belief that we are in Quadrant I, we do justice to the priorities and expectations of others.

Time management seminars are held again and again (you can't manage time, only make the best of it!) that various tasks, such as telephone calls, are to be processed collectively. The background is that we are interrupted by every call in our work and then have to start over again. This not only costs time, but also energy. Not to mention the fact that in a meeting, five employees have to wait until the project manager has finished the call. Then it costs five times the working time.

IV Quadrant of Waste

Quadrant IV is reserved for activities that are neither urgent nor important. Of course we didn't lose never anything there, but often we are so carried away by our fights in quadrant I and III that we flee in quadrant IV to get some air again. And what do we find there?

Not necessarily restful things, because true recovery is a valuable quadrant II activity. The reading of light novels, the consumption of mindless television programmes or the gossiping of coffee in the office can be attributed to the waste of time in quadrant IV.

But quadrant IV is not for survival, it's disintegrating. When the initial joy of sweet idleness has vanished, one notices that in this quadrant one not only loses time, but also oneself.

It is therefore important to focus on the essentials, define goals and redefine oneself whenever the situation has changed.

A few hints for planning and organization are mentioned in the following chapters.

6.5.1 Written Planning and Time Scheduling

Not only the definition of objectives is done in writing, but also personal planning. Written fixation forces us to deal with the activity in a very concrete way. She can't be forgotten that way. In the meantime, there are a wide variety of methods available, both online and offline.

Everyone has to decide for himself how he wants to plan and with which exactness. In the past, a very minute scheduling was trained, for example 8:45-9:30 telephone calls, 9:30-10:00 handling mail. This is not the right method for everyone. Instead, for example, an employee prefers daily to-do lists and processes them. Everyone has to find out for himself.

But it does work, however, to optimize the work and get a grip on the thieves of time by means of deliberate task bundling.

If it is an overcrowded desk, there are only four ways to clear it. You can download the documents

- throw away,
- forward,
- file,
- edit.

If you don't leave yourself no alternative, the desk will empty itself visibly. Don't allow for a fifth possibility, if possible already during mail processing. There's nothing you need to put aside first.

Throwing Away

As a project manager, you must be able to say no and set priorities. This also applies to the papers on your desk. Throw away what can go away. Ask yourself, "When did we last use the records?" A less helpful question is whether you might need something again. The answer is always,"Yes."

Everything,

- what can be easily replaced (brochures, catalogues, magazines)
- what is of little value or interest to you and your supervisor (invitations to events that nobody goes to)
- what is for your information only
- what is unimportant to you and your supervisor

comes away. Again, the goal of the project is in demand.

Forwarding

Forwarding is almost as practical as using the wastebasket. You get the stacks from your desk with little effort. Distinguish between documents that fall within your area of responsibility and those that are addressed to you, but do not need to be processed by yourself. Send these documents to the appropriate person or department as soon as possible. If possible, check with the tasks whether others could do them faster, better and cheaper.

Filing

Storage is the activity most people don't like to do. Documents, papers and notes are often piled up for weeks. In fact, filing costs a lot of time. The time required for filing depends on the type of filing, suspension file, folder, pendulum stitcher. The most common is the folder.

Today we often speak of the paperless office.

But if paper has to be collected, the NOGUCHI filing system (http://www.inpc.de/blog/2005/12/emergentes-ordnungssystem.html) is an excellent way to clear the table. In addition to the savings in personnel costs, there are other advantages: short access times, high flexibility due to the ability to adapt to the volume, identical filing in all levels, very good overview, ideal for "throwing in" the typeface from above, order according to importance.

Some criteria for a good and efficient filing are

- The "rediscovery rate" is high.
- Only the really important things are discarded.
- The time of filing per document is short.
- The costs (personnel, rooms, furniture) for the "maintenance" of the storage area are reasonable.

Editing

These are the tasks you have to do yourself. Put 4 baskets of mail in front of you on your desk and sort these works into the following categories

- Basket A: Reading material
- Basket B: Correspondence, mail, memos of conversations
- Basket C: Telephone calls
- Basket D: Other activities, such as travel expense reports, meeting preparation, travel planning.

Then you can structure your work by bundling tasks. similar tasks can be processed in a block. It is particularly easy to bundle routine tasks, such as mail processing, telephone conversations, meetings, appointments with customers, quotation preparation and invoice processing.

You will also get an overview of the routine tasks involved. Weekly or even monthly blocks of work can be set up for these tasks.

For many project managers, this sorting is not enough to optimize their day-to-day work. If not all the work that has to be completed on time, then these activities must also be prioritised.

Priorities are always determined according to the importance and not exclusively according to urgency, the deadlines. The inability to distinguish between urgent and important things is one of the biggest time thieves.

Here again the EISENHOWER principle is applied..

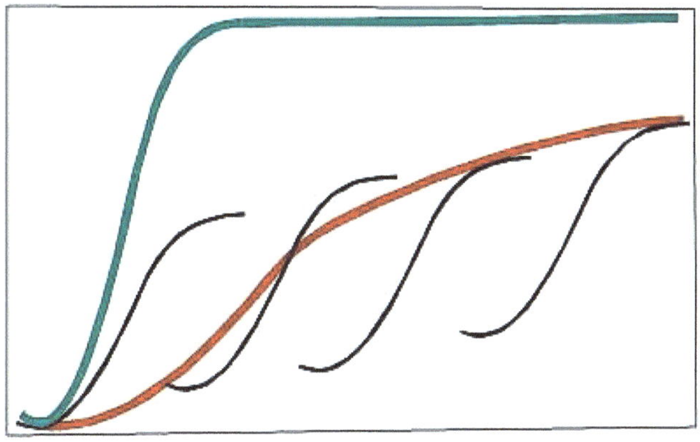

Figure 6.8: Work interruptions

Project plans can also be drawn up for small projects. Collecting and instantaneous in the correct order places, enables a clear structure of the pending task and allows less information to be lost.

6.6 Conflict Management

Tripping Hazard

If conflicts over time and resources are not resolved!

- Efficient scheduling or priority management means that the most important tasks and processes must be defined and arranged by the project manager or the team. The responsibility has to be clarified in advance..

Tripping Hazard

If the personal conflicts in the team are not solved!

- The project manager in particular should be able to identify and solve general causes of conflict, because certain conflicts in the team are completely normal processes that have to be run through time and again (team clock according to TUCKMAN (figure 6.9)).

6.6.1 Causes of Conflicts

Wherever people are closely related, there is a large number of conflict centers that, unresolved, block part of our available energy.

Because what is best for a department or group is not always ideal for everyone else. Nevertheless, such conflicts can lead to very creative solutions if handled correctly. Clever conflict resolution is an important skill of a successful project manager.

General causes of conflict

Conflicts arise from the fact that each person collects different experiences in the course of his or her life. His picture of the world is created in this way. It develops through his family, his values and attitudes, his environment, his culture and last but not least through the people he encounters every day. Every human being becomes so unique and has his own character. This results in quite individual expectations towards his fellow men and his environment.

Values and principles are therefore an integral part of every personality. Attacks on this very sensitive area act on the emotional level. When a person's values and principles are put into quest, his or her personality is always put into question.

This is a great potential for conflicts. Different expectations and attitudes meet each other. At the beginning of a relationship there is initially no understanding of the "otherness". This must be perceived and finally accepted. In the best case, it is understood. Only those who know themselves and appreciate others in their otherness can take into account others.

All people want to be treated with respect. But this consideration is already an intensive relationship between these two people. In conflict resolution methods, the relationships between two or more people are examined in greater detail.

In order to reduce the potential for conflict, it is very important not to draw conclusions from yourself to others. One's own point of view, one's own expectations, one's own way of tackling things is not the only one.

A sharpened ability to perceive others and their actions is an important prerequisite for the recognition of differences. It is also advantageous, as far as possible, to encounter foreign people without prejudice and empathic empathy. If one recognizes the reasons for a deviating behavior, it is much easier to understand and accept this. If living together is characterized by mutual understanding and acceptance, it becomes more pleasant. Disputes do not have to escalate and can be more constructive.

Motamedi /6.6/ mentions five reasons for conflict

- Digitisation,
- Own scripts of experience,
- Self-fulfilling prophecies,
- Divergent goals and interests,
- Values and principles.

Digitisation, i. e. the reduction of information through language, severely restricts the variety of ways in which we can express our thoughts and feelings. Communication is much more than the spoken or even written word. The complexity of communication is discussed in detail in chapter 9.

Every human being has scripts of experience. He stores his knowledge and experience in such scripts and activates them if necessary. For example, there are scripts for greetings, telephone calls or restaurant visits and determine the sequence of actions. In addition, they shape expectations and attitudes. The more complex everyday experiences are, the lower the capacity to absorb the actual events. The resulting inflexibility renders us incapable of acting and holds great potential for conflicts.

Self-fulfilling prophecies do the rest. According to Watzlawick /6.7/, these prophecies are both conscious and unconscious. If our behaviour is characterized by a certain attitude, people react accordingly. Thus, the self-fulfilling prophecy dictates behavior to the others, so to speak.

A successful conversation can only be conducted if both parties have the will to cooperate, to communicate with each other. The more similar the goals are, the easier it becomes. If the parties have divergent goals and interests, the discussion is problematic. Then it makes sense to first define a goal (9.2.1) in order to clarify to what extent the respective goals are harmonious. After that, the interests connected with the goal can be explained and clarified.

Project-Specific Causes

In addition to the manifold possibilities in the interpersonal area, there are a lot of conflict causes in the team during team development. This process is described as a sequence that takes place in different phases /6.4/. During this process, the team development clock (figure 6.9), which accompanies processes in the team in their individual phases, runs.

I. Orientation Phase or Forming

The first meeting takes place in the orientation phase or in the forming of team development. At this point in time, the roles and different behavioural patterns expected from a job or position are not yet defined. Uncertainty still prevails in this respect, the behaviour towards each other is polite. In the process, each team member finds his or her position and gets to know his or her colleagues better. If everyone knows how to judge the others and has found their place, this phase is over.

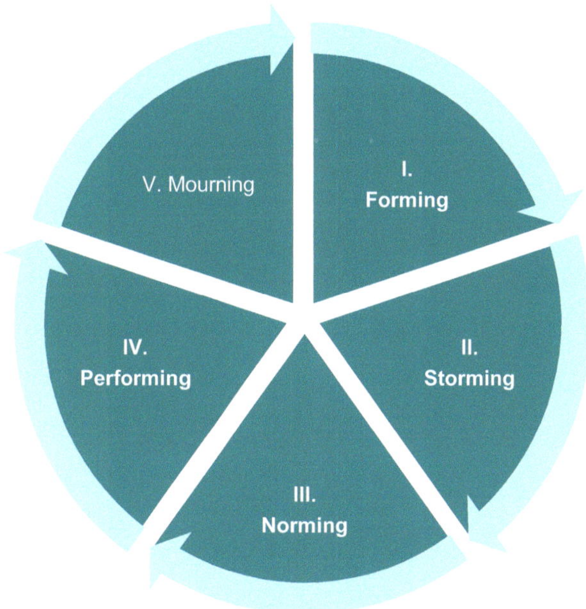

Figure 6.9 The Team Development Clock According to Tuckman /6.9/

II. Conflict or Delimitation Phase or Storming

During the conflict or delimitation phase or storming, the team members sometimes find that the agreements made first have to be changed because they are not feasible. It is also possible that individual members may try to gain too much influence or an inappropriate position. This automatically leads to conflicts. In this phase, the project manager is also a participant in this process.

III. Cooperation or Clarification Phase or Norming

In the cooperation or clarification phase or in norming, the team members make constructive appointments and behavioural decisions.

IV. Work Phase or Performing

During the working or performing phase, the team works harmonically with each other.

V Farewell Phase or Mourning

The team dissolves in the farewell phase or mooring.

Knowledge of this process is important so that the storming or conflict phase in particular is not hindered or interrupted. If conflicts are suppressed here, these can swarm on and disrupt teamwork under certain circumstances. Consequently, it is up to the project manager to control these processes in such a way that the phases are as short as possible, but as long as necessary, so that the productive phase can be reached as quickly as possible.

There are many reasons for the conflicts in the storming. They can arise between powerful and inefficient team members, as the difference in performance is sometimes seen as a legitimacy to enjoy advantages over others.

Furthermore, conflicts of authority and power can arise, especially between the project leader and individual team members. ranking question also offer potential for conflict, which means that people who have been part of a team for a long time enjoy a special rank and therefore new members must first pass a kind of trial in order to be regarded as equal or full members.

Today, employees are often isolated from their daily work routine, for example in congress hotels, and are brought together externally by coaches through appropriate exercises to form a team. By doing so, one hopes for a shorter storming.

A too little or too bad leadership of the team by the project manager also leads to conflicts. If the manager does not fill his or her roles sufficiently, a leadership vacuum /6.10/ is created. A person from the team will take over the substitute leadership without legitimation.

Poor leadership with an opaque information policy, special rights, inconsistency or conflict avoidance causes stress in the team. Different levels of information and information deficits are also a common cause of annoyance in projects. If information arrives too late, incomplete, distorted, false or not at all, it harms the project and leads to disputes between the parties involved.

Over- and underload can also be a source of stresses. The consequences of overstraining quickly become apparent, while understraining leads to frustration and boredom, and consequently to a preoccupation with other issues or even destructive behaviour.

Envy and rivalry also offer conflict potential, since under these circumstances personal interests and not the project goals are in the foreground.

The project manager is responsible for supporting the phase sequence, so that the team can get to know each other as quickly as possible and quickly reach the productive phase.

The previously described phase sequence can be found in this way or similarly for all projects, it takes place automatically. But it never happens the same. Each team is made up of individuals, which is why the process is individual for each team. No phase can be prevented or shortened, supported or hindered. Some even believe that if a phase is skipped, it must be made up for at a later point in time. The same happens with a so-called team shake, i. e. a change in the team structure, for example when a team member joins or leaves the team. Then the phase sequence starts again from the beginning.

6.6.2 Conflict Resolution Methods

There are always several methods of resolving conflicts. All methods are based on basic relationship patterns that depend on how the people relate to each other. The various possible solutions are illustrated in figure 6.10. Conflicts can be resolved in "my way","your way" or "our way". This results in different strategies.

Nearly every strategy for resolving a conflict is based on an interplay of willingness to assert and cooperation. This interaction is often impaired by emotions, which makes it sometimes difficult to find a satisfactory outcome for all parties. If one notices that a discussion is strongly influenced by the personal emotions of one person or another, the discussion should be postponed for a while to be resumed later.

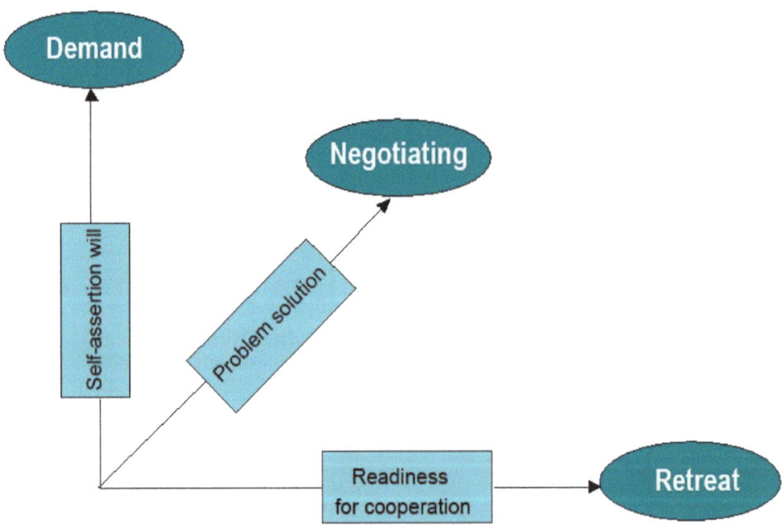

Figure 6.10: The Different Strategies for Conflict Resolution

In the meantime, one can have one-to-one discussions with the respective parties in order to gain a better understanding of the background. Here too, the tools from chapter 9 can be helpful.

Willingness to Assert Oneself and to Cooperate

Most people tend to take hardened positions when they are very concerned about a matter and when they are sure of their claim. They become compliant when they feel less influential than their counterparts.

Most people are cooperative when they respect the other party, when the relationship with the other party is important, and when they depend on the other party to enforce a decision.

The interaction of these two factors is reflected in conflict resolution strategies.

Demanding and Giving In

Claims are generally presumptuous and uncooperative. They show the certainty of victory and the importance of the topic for the demanding party. However, they can also be characterized by the disinterest in the relationship and the absence of dependence on the other party.

Simply giving in is a very cooperative tactic, but generally shows little willingness to assert. Either the subject matter is not important to the person concerned, or knowledge of it is too little to exist in a dispute. But it would also be possible to simply leave the other person's will in order to establish or maintain a good relationship with him.

The problem-oriented solution strategy shows both a strong willingness to assert oneself and a great willingness to cooperate. It signals that this is an important issue, but that a good ongoing relationship with the other side is also desirable.

Negotiating

Negotiation is filled with both factors, the willingness to assert and the willingness to cooperate. Negotiation begins when an important issue is raised by peer-to-peer partners. Each side must be prepared to yield a piece to achieve a constructive solution. Negotiating is also a good substitute if a joint solution to the problem has not yielded a clear result.

Each of these strategies has its own appropriate scope. Often one restricts oneself to an "educated" strategy for conflict resolution and applies it in every situation. The trained conflict partner recognizes, however, which strategy is appropriate in which situation and thus has a greater chance of success through the choice.

These behavioural patterns can be found in the different conflict resolution methods. Whether it is the inclusion of another person who acts as a mediator, as in mediation, or whether both parties try to reach a compromise using the HARVARD Method. It is always about finding a solution to a problem that both parties can live with.

In order to achieve the desired success, however, the cooperative relationship between the parties is always important.

However, this negotiation strategy will only be successful if both parties are really willing to cooperate. If they are hardened positions or naysayers of one or both parties, negotiation will hardly or only painstakingly lead to success.

6.7 Test Your Progress

Now you can check your knowledge in table 6.7.

Table 6.10: Chapter 6 Test: What statements are correct or false?

No.	Question	correct	false
1	Negotiating materials and services is not the task of the project manager.	☐	☐
2	Control is the key activity during the implementation phase.	☐	☐
3	Specifications, schedules and budgets developed during the planning phase become the yardsticks for measuring the progress of the project.	☐	☐
4	Control point detection does not normally save time.	☐	☐
5	A project control chart summarizes information about the quality dimension of a project.	☐	☐
6	A milestone chart is not detailed enough for project planning.	☐	☐
7	Budget information cannot be displayed as a line chart for control purposes.	☐	☐
8	Tests provide important information about the quality dimension of a project.	☐	☐

Continuation Table 6.10

9	Personal inspections are the most ineffective way to monitor project progress.	☐	☐

10	Invoice checks are only carried out at the end of the project to check the budget.	☐	☐
11	A progress review requires good listening.	☐	☐
12	Any work not conforming to the specifications must be repeated.	☐	☐
13	Lost time can often be made up for in later steps.	☐	☐
14	Using more resources can save customers money in the long run.	☐	☐
15	It is never acceptable to renegotiate or revise the conditions of a project after it has been started.	☐	☐
16	Effective feedback focuses on actions, results and responses to an event.	☐	☐
17	It is not important to communicate clearly when communicating feedback, because the person you are talking to already knows what it is about.	☐	☐
18	Successful negotiations require a power advantage over the other party.	☐	☐
19	In negotiations, the one who offers attractive alternatives has a good position.	☐	☐

Continuation Table 6.10

20	Giving in is never a good strategy for resolving conflicts.	☐	☐

1	2	3	4	5	6	7	8	9	10	11	12

f	r	r	f	f	r	f	r	f	f	r	f

13	14	15	16	17	18	19	20
r	r	f	r	f	f	r	f

7 Project Completion

The most valuable thing you can do is make mistakes - you don't learn anything when you're perfect.

Adam Osborne

7.1 Bringing the Project to a Successful Completion

The goal of the project management is that the client accepts the result of the project and that the company makes profits with the implementation of the project.

The first point means that the client accepts the finished project which meets the quality specifications. In order to guarantee the smooth running of this evaluation, clear performance criteria were defined by the client and project manager from the beginning of the project. These can be adapted and modified during the course of the project, but all changes must be recorded in the contract with their impact on the schedule and budget, or they must be agreed with the client.

In all performance criteria, care should be taken to ensure that they are measurable and concrete, i. e. objectively comprehensible. Of course, this is not always exactly possible, but should always be aimed at in order to avoid problems with the performance decrease later on. Another important criterion is the purpose of the project.

The project does not have to be completed with the delivery to the client. The technical documentation, such as operating instructions, plans and the final report, generally follows after delivery or completion. Likewise, training of the operating personnel usually takes place after delivery or completion. Performance tests are also performed after the end of the warranty period, for example, a continuous test or a trial period during which a special warranty is specified.

After completion of a project, the project team members have to be integrated into new projects or tasks. Excess material and equipment must be disposed of and used equipment must be returned to its original use. The place of installation or construction site must be disbanded.

An important last step after the completion of a project is an evaluation of the project. Through this review of the overall project, the experiences and insights gained will be considered again in order to apply them to future projects. In general, the review takes place as part of a group discussion with the most important project members.

7.2 Checklists for Project Completion

Often the project manager is already busy with another new project and has to finish the old one. Table 7.1 shows a checklist for the completion of a project, in which the various activities can be checked by the project manager at the end of the project or delegated for further execution.

7.3 Questions about Project Evaluation

The following questions can help to draw benefit for future projects from the experience gained in the completed project. The more careful this is done, the better it will be for the future project teams.

Table 7.1: Completion list

1	Check the project result for proper functioning.	☐
2	Write operating instructions.	☐
3	Making plans.	☐

Continuation Table 7.1

4	Deliver project results to clients.	☐
5	Train the client's personnel in the use of the product or facility.	☐
6	Transfer project team members.	☐
7	Dispose of excess equipment and material.	☐
8	Return facilities and premises that are no longer needed.	☐
9	Summarize the problems encountered and the solutions used.	☐
10	Keep track of technical progress.	☐
11	Summarise recommendations for future development.	☐
12	Summarise what has been learned through cooperation.	☐
13	Write performance reports on the members of the project team.	☐
14	Give all employees feedback on their performance.	☐
15	Carry out the final inspection.	☐
16	Write final report.	☐
17	Project review with the core employees.	☐
18	Declare the project complete.	☐

Therefore, it is certainly useful to answer these questions in written form and pass them on to the management. However, since the project manager also has to "admit errors", this open process only functions if there is a relationship of trust between the individual parties.

- Did the project meet the planned timetable?
- Will this result in changes to the future schedule? Which one?
- Could the planned budget be met?
- Will this result in changes to the future budget allocation? Which one?

Table 7.2: Chapter 7 Test: What statements are Correct or False?

No.	Question	correct	false
1	The project manager's work ends with the handover of the project result to the client.	☐	☐
2	The client must agree to the performance criteria before starting the project.	☐	☐
3	Objective, measurable performance criteria are easy to develop.	☐	☐
4	Writing operating instructions and training the client's personnel in the operation of the project result are part of the final phase.	☐	☐
5	Normally, at the end of a project you don't have to worry about excess equipment or material.	☐	☐
6	Assigning new tasks to project staff is one of the final steps in completing a project.	☐	☐

Continuation Table 7.2

7	The project manager's responsibility ends with the completion of a final review and the writing of a final	☐	☐

	report.		
8	It is of little use after the completion of the project to waste time evaluating the experience gained.	☐	☐
9	The implementation of the project often results in technical progress, which is worthy of being communicated to other departments or the company management.	☐	☐
10	If the project experience gained is not recorded, it is usually lost and cannot be passed on to other project managers.	☐	☐

1	**2**	**3**	**4**	**5**	**6**	**7**	**8**	**9**	**10**
f	r	f	f	f	r	r	f	r	r

Tasks

- Is the project manager's work finished when the project result is handed over to the client?
- Which works are part of the final phase?
- Name three instruments of project monitoring.
- Indicate five questions to be discussed in the framework of the project evaluation.

8 Project Management Programs

Tripping Hazard

If unsuitable systems and processes for project planning and control are available!

- The tools must be adapted to the project size (paper versus project management software versus scrum board) and must be suitable.

With the appropriate project management software, for example MS Project®, which is representative of many others, comprehensive project management and an efficient exchange of information is achieved. The project manager moves from individual planning to company-wide planning.

In the following, the program Microsoft Project 98® (MS Project®), or the latest version, is presented in more detail. However, the author expressly points out that it is not the only available for project management programs on the mark program et.

Due to the network capability, the requirements for improved communication and cooperation of many projects are fulfilled. This means in detail that the project manager has good control over the project plans, is able

- to maintain efficient communication within the project team,
- has a powerful tool for creating user-defined planning solutions at hand.

In the context of this book only a short description of the software can be given. In the relevant literature, for example manuals on the individual programs and articles on the latest versions, interested users will find further information /8.1/.

8.1 Control over Project Plans

With MS Project® the project manager can create and pass on schedules and project plans.

MS Project® has a planning module that allows you to

- plans and supervises the projects step by step. This gives the project manager an overview of all details in the project.
- by functions, such as interrupting individual steps (operations) during work, changing the availability of resources or changing cost rates during a project, that reflect the project reality well.

8.1.1 Project Scheduling and Monitoring

Process Interruption

The project manager can interrupt operations. It first clicks on the corresponding button on the toolbar and then divides the process into smaller segments by dragging the mouse pointer. Then he can divide an operation into any number of partial operations, whereby the duration of the interruptions does not matter (figure 8.1).

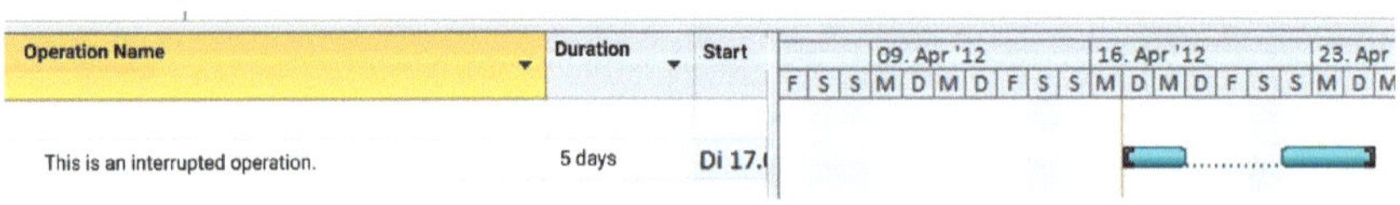

Figure 8.1: Interruption of Operation by Dragging the Mouse Pointer over the Operation Bar in the Bar Chart

Activity-Based Scheduling

In some cases, the project manager would like to concentrate all available resources on certain operations so that the work is completed as quickly as possible. For example, if a worker was put down for painting a wall, a second painter could be used (figure 8.2) to complete the wall in half the time originally set.

With MS Project®, the duration of an operation is automatically extended or shortened if an employee is additionally assigned or deducted from an operation. The change will take over. However, the total workload for the operation does not change.

Variable Resource Sets

Changes in wages and salaries can be made over time so that the project manager can closely monitor costs. For each resource there is a table (figure 8.3) that contains the cost records with the time attributes. These cost rates are used to calculate project costs when the resource performs work for the project.

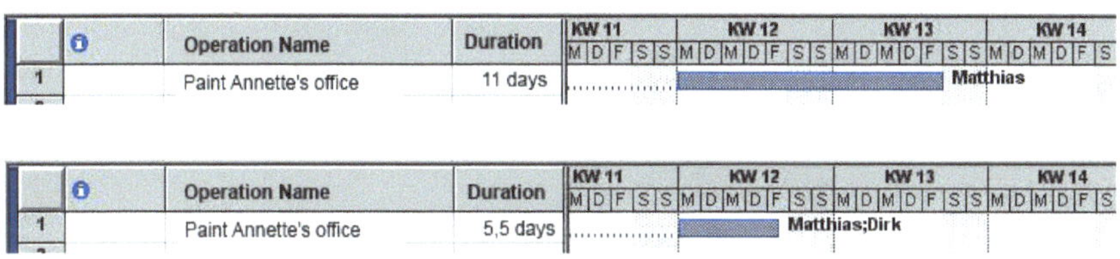

Figure 8.2: Activity-Based Scheduling

The project costs will be charged at the rate applicable when the work is carried out. The project manager can write changes to the remuneration rates in this table either by specifying a new rate directly or by entering a higher or lower percentage. The program then calculates the new block automatically.

In addition, up to five different cost rate tables can be used to assign different cost rates for different operations to a resource. For example, a resource can have an hourly rate of € 53 for one operation and an hourly rate of € 85 for a higher value operation. The project manager specifies the cost rate by assigning the corresponding cost rate table.

The due date of the costs can also be specified in three different ways: proportionately, at the beginning or at the end of the allocation.

8.1.2 Management of Multiple Projects

There is no upper limit for the number of projects. This allows many projects to be controlled side by side. In addition, there are some useful tools for coordinating resources and processes between projects. The functions for managing multiple projects in MS Project® are shown below with some examples.

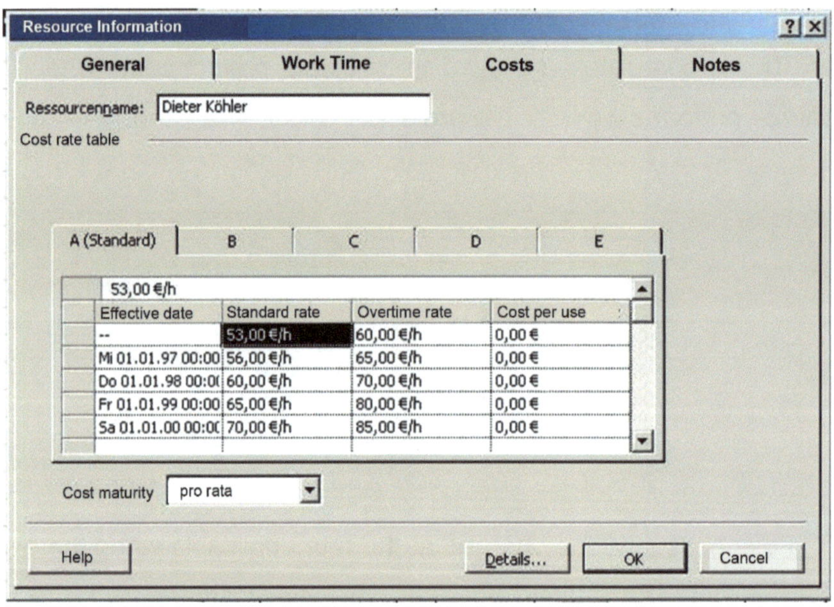

Figure 8.3: Enter the Cost Records on the Tab Page

Inserted Projects

A function allows the project manager to insert complete existing projects at any level into the structure of a parent project. The inserted project can then be hidden or hidden to view all details of the project in the context of the parent project plan.

Either a copy of the subproject is inserted, which represents the status at a certain point in time. Or, a link is established between the parent and subordinate project. This means that changes made in one of the two plans are automatically displayed in the other (figure 8.4). The overall project illustrated in the example consists of three separate projects: the plan for the online catalogue, the infrastructure for the online catalogue and the plan for the printed autumn catalogue.

The inserted projects can be found at any level of structure and have exactly the same functions as the independent individual projects.

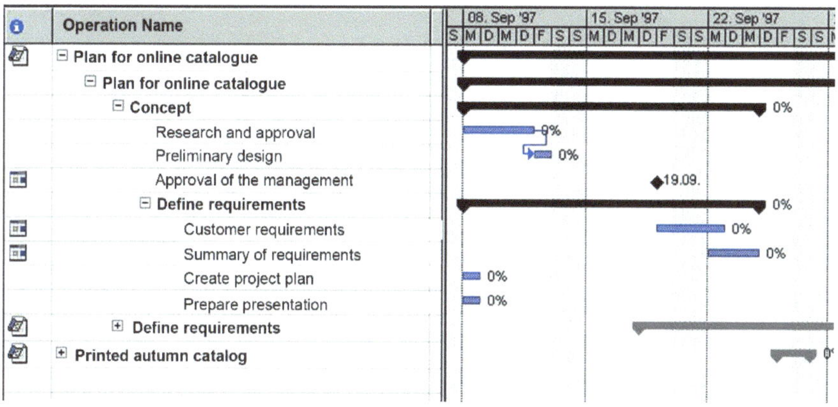

Figure 8.4: Overall Project Consisting of Three Separate Projects

To insert a project, simply click on the line in which the project is to be inserted and then on “Insert project” in the menu.

Another feature of the inserted projects is that cross-project links are displayed as normal dependency lines. The external operations do not appear, which hides redundant operation information.

Ressourcenpool

For the project manager, the task of managing multiple schedules is very complex because resources are often shared with other projects.

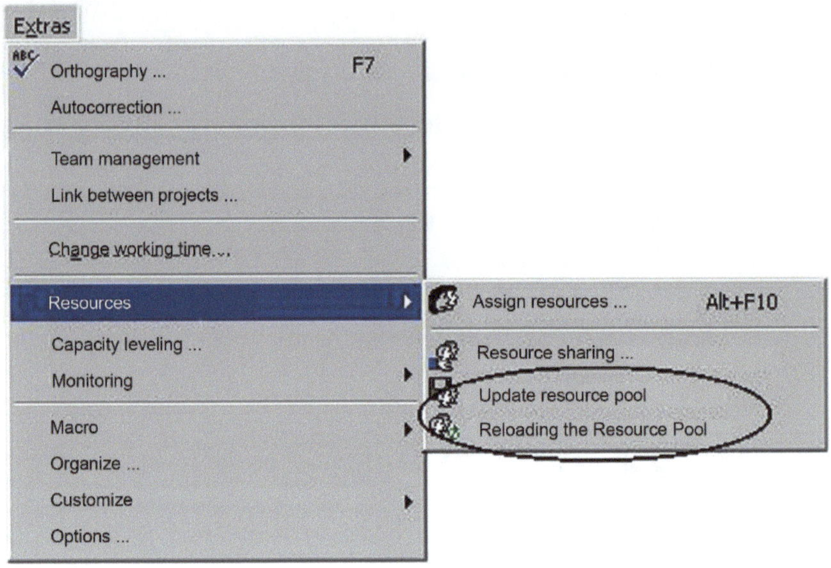

Figure 8.5: Update Resource Pool

The program allows the project manager to open a read-only snapshot of the resource pool so that he can see the status of shared resources. If he wants to apply changes made by other project managers, he can use the "Relaunch pool" option to download the latest changes. If he himself makes updates that affect the resources in the pool, he can inform the other project managers about these changes with the "Update pool" function (figure 8.5). The resource pool function gives project managers the control and flexibility they need in an environment with multiple projects running simultaneously.

8.1.3 Analysis Tools

AutoFilter

The AutoFilter function can use drop-down lists at the top of each column to use a view to interactively filter out certain operations, resources or assignments in projects (figure 8.6).

Custom Filters

The project manager can also use dynamic filters with natural language usage and also make a user-defined selection using the dropdown list to create his own filters with several conditions and save them for the future application. This allows them to concentrate on specific entries in the project plan.

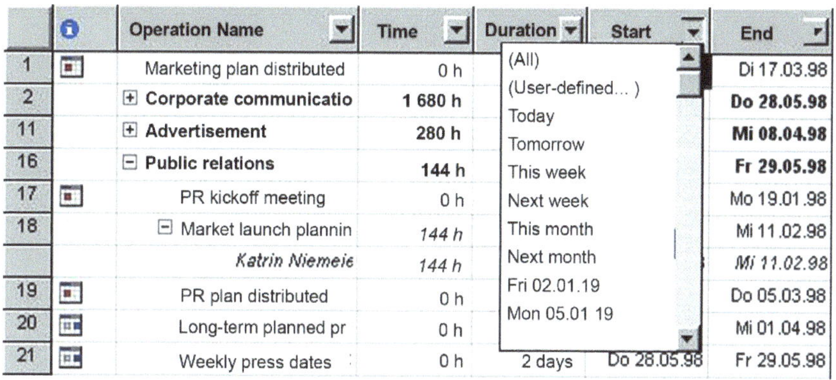

Figure 8.6: Using the AutoFilter Function

Due Dates of the Fixed Costs

The project manager himself determines when the project costs are due. They can choose whether the fixed costs are incurred at the beginning, end or proportionately during the process (figure 8.7).

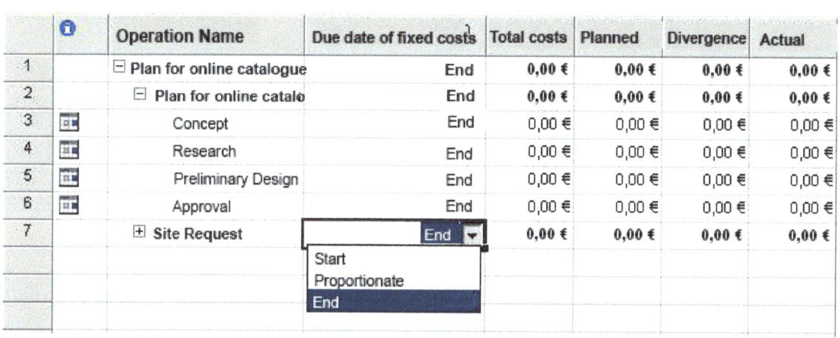

Figure 8.7: Determination of the Due Date of the Fixed Costs

Transaction Types

The standard equation for calculating the operation processing and duration is work = duration time units. However, the project manager can control which elements should be fixed and which are variable, depending on how the planning is carried out in the company. The program supports three types of operations: fixed duration, fixed unit or fixed work.

Appointment Planning from the Finish Date

Frequently necessary processing of scheduling from the final date is possible in the program.

Capacity Alignment

The program contains a number of features that the project manager can use to control how the algorithm works.

The project manager can change the setting in such a way that the overload is regulated, for example, at weekly intervals.

The project manager can carry out capacity leveling for the entire project or only for a certain period of time. In capacity matching, individual assignments can be moved so that only the resources affected by the overload are moved. However, if all resources have to work simultaneously, this can also be set.

You can use this function to plan capacity leveling around operations with an overload. For example, a week-long process starting on monday can be interrupted for a necessary meeting on wednesday.

Setting "Week Start On" and Option for Fiscal Year

All weekdays can be defined as the beginning of the week (figure 8.8). The definition of the fiscal year start can also be set there.

"Hard“ and „Soft“ Limitations

The project manager chooses the way in which the program handles operations that violate restrictions. An option is available on the schedule tab in the options dialog box of the tools menu. You can deactivate the option consider operations always their constraint dates to define soft constraints, where operations can be planned beyond the constraint dates.

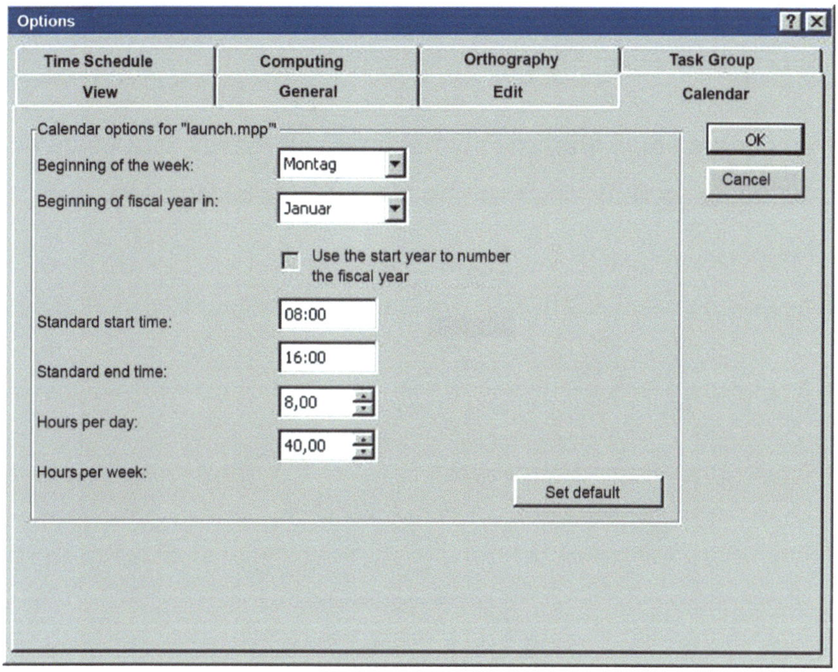

Figure 8.8: Determination of the Beginning of the Week and the Fiscal Year

This setting affects the constraints MAA (must begin on), MEA (must end on), ENSA (end not later than) and ANSA (beginning not later than) if the schedule starts from the start date, but also the constraints MAA, MEA, ENFA (end not earlier than) and ANFA (beginning not sooner than) if the schedule starts from the end date.

Multiple Critical Paths (Critical Path)

The Calculate Multiple Critical Paths option is another option on the calculation tab of the tools menu's options dialog box. It sets the last finish date of an operation to the earliest finish date as long as the operation is not restricted by successors or restrictions. In this way, several critical paths of the project can be calculated. This is particularly important if the project manager wants to merge several project plans into a single project file and see the critical paths for each individual project.

Indicators

Indicators are small graphic information about events, resources or allocations (figure 8.9). They can be viewed by clicking on the indicator column i.

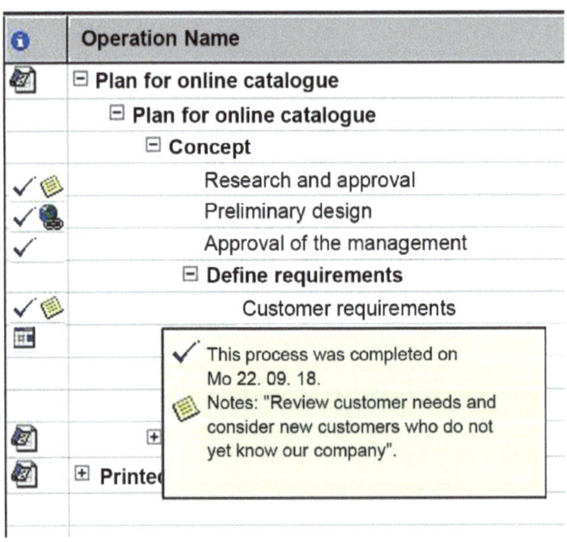

Figure 8.9: Quick Overview through Indicators

Line of Progress

The progress line is an additional function of the bar chart (GANTT chart) that allows visual representation of the project status.

This line connects the respective progress of the operations with a vertical line on the selected date (figure 8.10). This results in tips pointing left and right for the individual steps. The work that falls behind the schedule shows to the left, which is ahead of the schedule, to the right. This enables the project manager to see at a glance the overall status of the project and the extent of deviations from the plan.

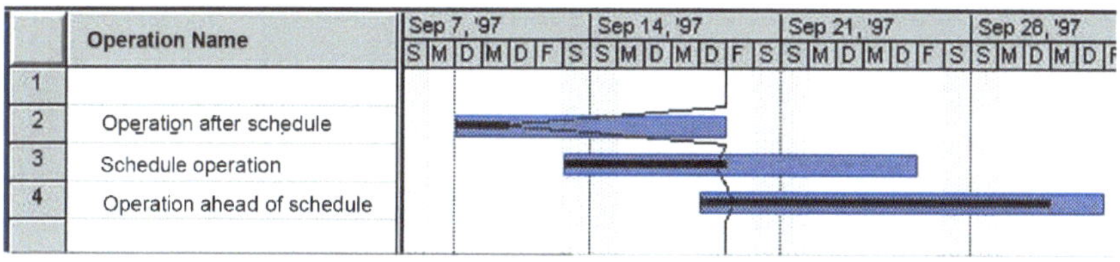

Figure 8.10: Line of Progress

8.2 Communication within the Project Team

The project manager must be able to work efficiently with very different people. To do this, it must be able to inform project team members and managers at all levels about the project status at any time.

Successful cooperation depends to a large extent on the clear and efficient dissemination of this information. The project team can share information via email, the Internet and the intranet to collaborate with others.

8.2.1 Presentation Features

Reports

Most project plans are published as reports or stored as project files on a server for distribution to other project managers.

MS Project® has a library of predefined reports that the project manager can print out immediately. There are also reports that can be used to trace cross-project dependencies. Here too, the project manager can create his own reports using the combination of views, tables and filters.

With the aid of the GANTT chart wizard, the project manager can quickly create professionally designed reports.

The design of the headers and footers, for example with company logos or a formatted text, can be created easily (figure 8.11).

Alignment of Header and Data Fields

The project manager can choose the alignment of the column headers and data independently of each other.

Formatted Notes

The text and information in the memo fields can be formatted to highlight any important points. Other objects, such as documents (figure 8.12), can be included to provide additional information.

Time Ccale

It is also possible to display two special time scales for the project information in a project, e. g. semi-annually and in thirds of a month (figure 8.13).

8.2.2 Teamwork and Monitoring

The project manager can delegate and monitor project processes via email. A workgroup function allows access to a desktop information manager and the internet.

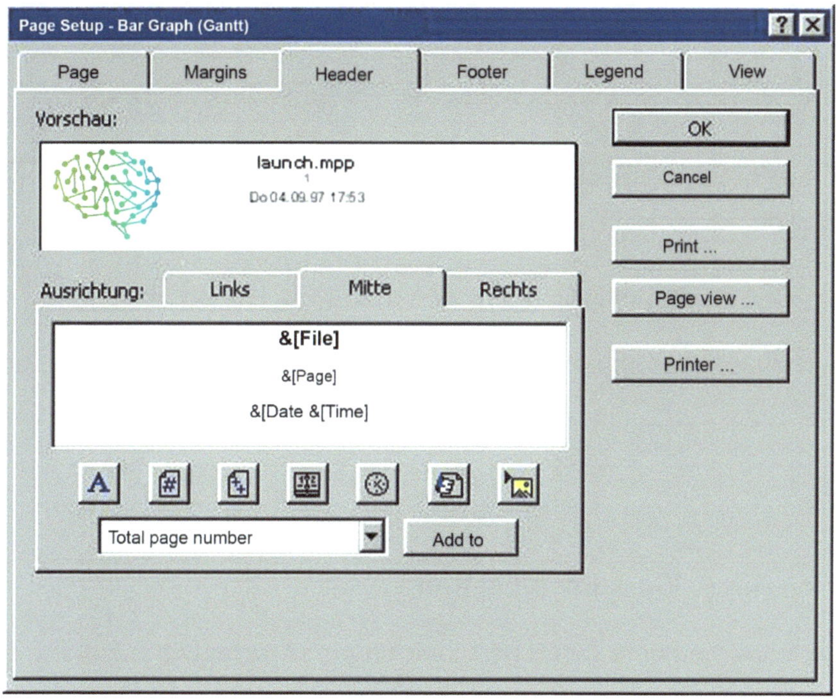

Figure 8.11: Freely Configurable Headers and Footers

Project managers can define reminders for certain project processes and store them in a desktop information manager. For example, the project manager can define a reminder that displays "x days" before the end of an important project phase of the project plan.

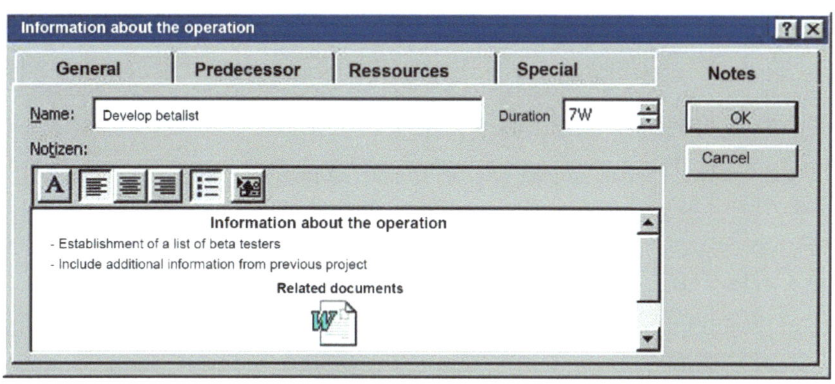

Figure 8.12: Project Notes and Integration of Other Objects

a)

b)

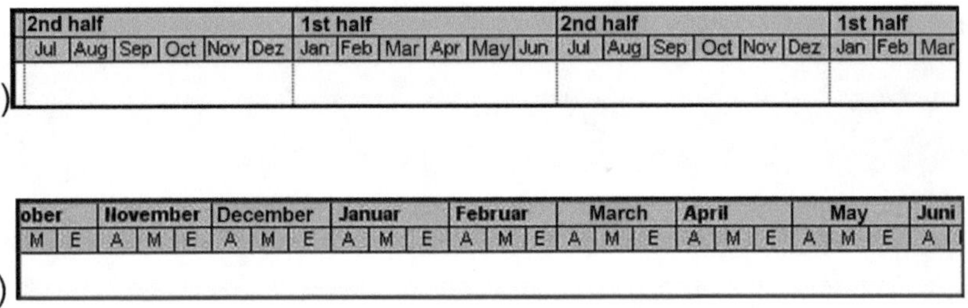

Figure 8.13: Time Scale; a) in months; b) A, M, E (start, middle, end of Month)

Email-Based Workgroup Feature

The team members can report their progress at regular intervals. The customizable email messages contain fields that can include time-relevant plan and current work data.

MS Project® is integrated with a desktop information manager task list. In this way, team members can centrally manage all processes sorted by project name.

Team members can select the team status report from the a desktop information manager tasks menu and save this status report in their inbox until they want to submit it to the project manager.

Only a small selection of possibilities offered by the MS Project® project management program is shown here. In the relevant manuals you will find further detailed descriptions of the possibilities.

8.3 Classical Versus Agile Project Management

Tripping Hazard

When the ScrumMaster becomes a project manager!

- The ScrumMaster is a role which should be taken by several team members alternately. To do this, the team members must be qualified.

Tripping Hazard

If individual team members do not want to take responsibility for their tasks!

- The teams must be able to lead themselves. This requires individual team members to take personal responsibility.

To do this, the team members have to be motivated, for example with the Instant Influence Method.

Tripping Hazard

When planning and design are not measurable!

- Scrum introduces supporting features for adequate business and subsequent requirements analysis.

For fear of overlooking important functions in the project, a comprehensive functional specification is developed. This is often overburdened with unimportant factors.

On the one hand, this leads to an extension of the project period and on the other hand to unnecessarily high costs for the client.

During project planning, it is often difficult to foresee which of the functions or interfaces will actually be required at the time of project completion.

For the continuous updating of the contract, the so-called change management during the implementation process is agreed upon. Thus, every flexibilization represents a foreign body that interferes with, hinders and costs resources in the ideally linear course of the project. For the client this means additional unplanned costs /8.1/.

In contrast, agile project management is a process. The itera-tiv leads to the solution. It is generally regarded as customer-oriented, fast and cost-effective.

At the beginning of the project a rough project vision takes place as project definition, which develops into a detailed project goal through repeated planning and implementation phases.

Developers and clients work closely together, discuss all problems and solutions with each other and make the necessary decisions together. It is intended that the subject matter of the contract will be further developed and changed.

This agility thus promises a needs-oriented and contemporary project result.

Absolute agility cannot be guaranteed, however, as too many human resources would be tied up.

Costs also remain intransparent because it is not clear what exactly should be developed in the project.

In this way, it is not possible to determine in advance what the development costs in the end.

The "agile fixed price" tries to create cost transparency.

There are two methods of agile project management.

8.3.1 Scrum Method /8.2/

Scrum is a lightweight planning tool for the development of products, projects and systems that has been used successfully for more than 10 years. It supports the use of agile and lean principles with the aim of being able to react quickly and flexibly to changing requirements and to deliver high quality with an increased return on investment. It is meaningfully complemented by agile development practices such as test-driven development and continuous integration.

The scrum method subdivides different roles within a scrum team, which can change from sprint to sprint (figure 8.14). It is the ProductOwner, the ScrumMaster and the team that performs the actual project task.

ProductOwner

The ProductOwner is the interface between the development team and the customer. He conducts consultations with the customers and defines all product requirements and customer wishes in the ProductBacklog.

ProductBacklog

The ProductBacklog is a prioritized function list, with short descriptions of all functions that are required by the product. With the first application it is not necessary to start a project, from which all requirements are documented at the beginning.

In the so-called SprintPlanning subtasks part of the ProductBacklog are extracted, defined as a task package and stored in the SprintBacklog.

But scrum does not specify a structure for the backlogs and says nothing about the structure of the entries.

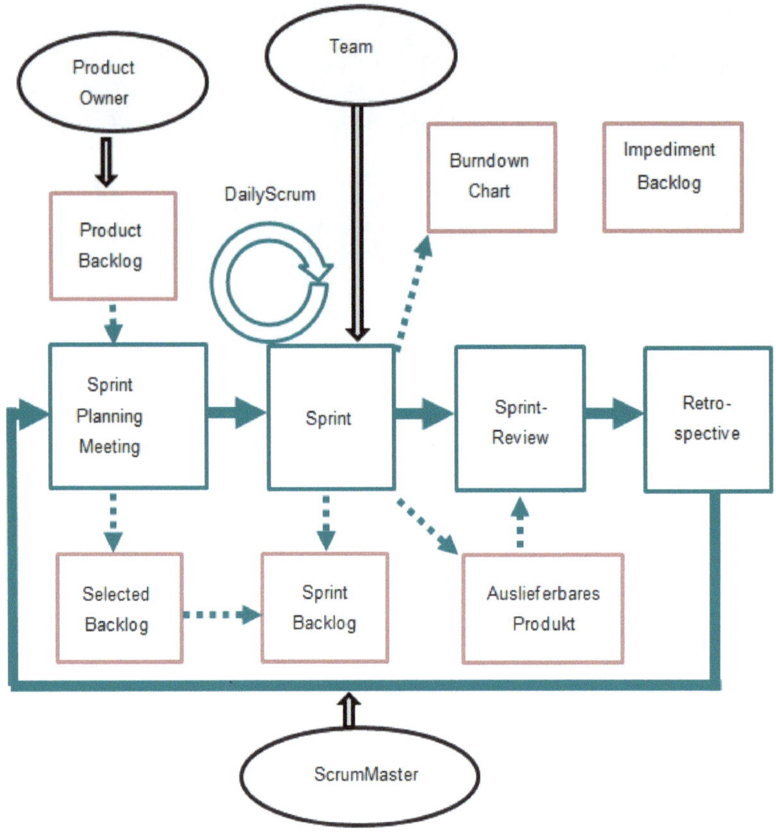

Figure 8.14: ScrumProcess

Sprint

The SprintBacklog is then processed in the scrum as a sprint defined processing block. Ideally, a sprint should ideally have the same length of time and lie within two to four weeks.

The development of the product takes place in consecutive sprints.

DailyScrums take place during the sprint, short meetings where team members report on their progress.

During a sprint, the team is able to deal with the tasks without any disruption, such as customer changes.

For small teams, four weeks is often too long for a sprint. Shorter cycles are advantageous.

ScrumMaster

The ScrumMaster has an essential role in the execution of scrum. Unlike a project manager, he does not take responsibility for the planning and control of a project, but rather supports and motivates the team as coach and moderator in self-organized work. It promotes business success by optimizing cooperation between customers and development.

The ScrumMaster provides additional support by protecting the team from external influences.

Furthermore, he ensures that the ScrumProcess is adhered to and helps with the processing of obstacles (ImpedimentBacklog) without solving the problems himself.

SprintReview

At the end of each sprint the SprintReview takes place, here the team presents to the ProductOwner what it did in the sprint. The new requirements and changes will also be discussed and entered into the ProductBacklog by the ProductOwner.

The SprintRetrospective, in which good and bad things are discussed retrospectively during project implementation, initiates a continuous improvement process /8.3/, /8.4/, /8.5/.

BurndownChart

A BurndownChart is a graphical representation of the leaded effort in a project. The backlog is usually displayed on the vertical axis, the time on the horizontal axis.

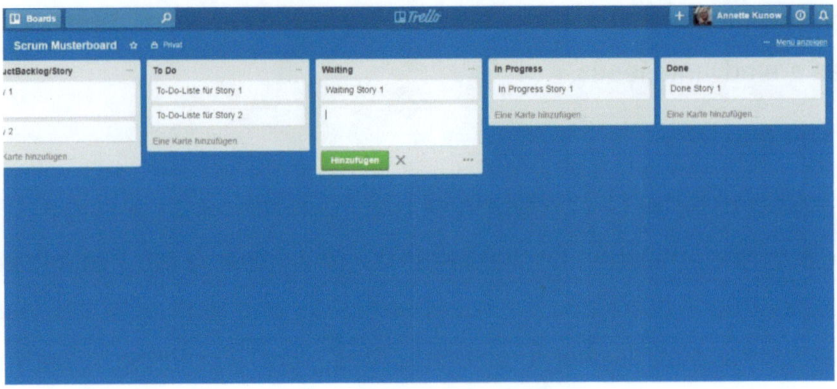

Figure 8.15: ScrumBoard with Trello

ScrumBoard

A ScrumBoard is a board created with a ScrumTool, for example the software Trello (figure 8.15 https://trello.com/). ScrumBoards are for teams that plan their work in sprints. A ScrumBoard is visible in planning and work mode.

8.3.2 Design Thinking

Design thinking is a particularly creative method of project management and focuses on the customer's needs. The entire process is a dynamic process. The project steps are continuously changed in the course of the ongoing project.

Design thinking involves seven steps. Within these seven steps, problems can be summarized, the correct questions asked, more ideas put together and the best answers selected.

This process can be repeated for individual steps in order to achieve the optimal project result by this iterative procedure /8.6/, /8.7/.

Step 1: Define

Here the formulation of specific questions is worked out in order to determine the needs and challenges of the project.

Step 2: Research

By means of a subsequent, intensive research and field observations, information about the product is collected, which can be extended by experts from the respective project areas, for example. Of course, personal experience can also be brought into the project.

Step 3: Finding

Brainstorming methods are used to create ideas for the project.

It is important to achieve as many creative solutions as possible. Whether such a new idea makes sense and/or is economically viable must not play a role in this phase. Any criticism must be refrained from in order not to interrupt the process of creating ideas.

Now the comprehensive information collected in this way is focused, i. e. the information relevant to the project is bundled and applied. In addition to new constructive ideas, this also includes possible mistakes from previous projects.

Step 4: Building the Prototypes

Once a sufficient number of ideas have been found, simple prototypes are created to test them on the target group and to show the advantages and disadvantages of the product.

Step 5: Selecting

Finally, the insights gained from this step will be used to develop the final product. All changes to the solution proposals made by testing the prototypes are incorporated and/or the necessary corrections to the previous solution model are introduced.

Step 6: Implementing

The resulting solution is executed.

Step 7: Learning

Based on the experience gained, knowledge for new projects is developed.

8.3.3 Case Study of the Planning of a Trip with Scrum

The structure of the project "trip" corresponds to the classical approach in project management. In addition to the requirements and goals defined at the beginning, it contains a fixed schedule and project specifications. In contrast to the approach in agile project management, this project does not provide for continuous consultations with stakeholders (students, transfer and accommodation managers, etc.).

For this reason, the requirements and objectives are unchanged over the entire project period. However, the projector organisation provides for a one-off information event in which the variable travel locations can be presented and agreed upon by the students.

Furthermore, feedback from the students is possible at any time during the project by e-mail in order to express special wishes, suggestions and criticism. The feedback influences both the requirements and the objectives of the project, e. g. when the cultural interests of the participants change during the course of the project.

With regard to agile project management, further deadlines must be scheduled for an incremental or iterative approach over the entire period of time to be agreed with all stakeholders in order to adapt to changing requirements and objectives. These remain variable at the beginning of the project.

Ongoing coordination of project goals and progress would increase the likelihood of success and lead to greater satisfaction with the project results for all those involved.

However, there would also have to be a higher deployment of personnel from both sides.

Here you can book a free Strategy Session or write me if you like this book and have suggestions or questions.

Click here for free bonus material on the book.

Also visit my blog „Selbstführung & Produktivität in German“. I'll help you get better results.

9 Brainstorming and Problem Solving

Imagination is more important than knowledge.
Albert Einstein

Those who want to carry out successful projects need tools for brainstorming and problem solving. On the one hand, project managers have the opportunity to find good ideas even with abstract tasks. On the other hand, a professional communication helps them to react appropriately and quickly identify the causes of conflicts.

9.1 Creativity Methods

There are a whole range of tools for brainstorming, the creativity methods. Only the best known ones are listed here,

- Brainstorming,
- Mind-Mapping,
- MOSES method,
- LEODARDO principle
- ABC method according to BIRKENBIHL.

These methods are informal ways of working, tapping the creative potential of an individual or a group through free association of ideas.

Through the free flow of thoughts, far-reaching correlations of a term come about. The methods serve to open up as many aspects of a new, perhaps abstract topic as possible.

In the abstract concept of stress, for example, this ranges from the mechanical meaning to the mental state of the human being.

These multifaceted aspects allow the term to be considered comprehensively. With some practice these methods to develop and then structure large projects. In the following chapters, these methods are explained individually.

9.1.1 Brainstorming

Every business has to be treated artistically if it is to succeed safely and permanently and in a thoroughly expedient manner.

Novalis

This method is usually carried out in groups. When a group member expresses an idea, this impulse triggers associations and new ideas among the other group members, which can lead to further ideas and so on.

This efficient tool supports the concretization and analysis of new ideas as well as the derivation and development of further ideas. It also allows new ideas to be viewed from a variety of angles and perspectives. It provides a framework for capturing and linking new ideas and is also a platform on which these can be translated into plans, projects and activities.

In the lecture Project Management & Business Coaching a self-edited project is required instead of a written examination. Of course, the students first have "no idea" what to do at the beginning. Through a brainstorming session possible project ideas are collected, which are sorted according to project topics.

Brainstorming Procedure

- All ideas brought in by the group members are recorded, for example on a flip chart or with the help of note cards on a pinboard.
- The collected ideas are not assessed or evaluated at this time, except to clarify any misunderstandings and are not yet sorted.
- Even far- fetched ideas are openly accepted. The discussion on this will take place later, and then we can still delete it.
- Repetitions are okay and will not be checked or sorted out at this time. This could disrupt the process.

- Quantity comes before quality. The more ideas are recorded, the greater the possibility to find a useful one.
- The brainstorming session must not be ended too early. If the flood of ideas slows down, a short pause is useful. Afterwards the brainstorming can be resumed.
- Afterwards, the group carries out a joint review and structuring of the newly found ideas. Now we sort, judge and discard. This generally requires a moderator to lead the group. This role falls to the project manager.

The project manager is always also the moderator of the team. However, he rarely has a special training for this, which is why there are often problems in the application of creativity methods. For example, it is often not possible to unravel the results of a brainstorming session completely, especially if some time has passed since the session and the logical further development of the ideas is no longer clearly comprehensible.

Brainstorming sessions often show that the procedure is not applied correctly. Many moderators don't take the time to gather ideas without rating.

Immediate sorting is like censorship: Now the participants no longer associate themselves freely, but ask themselves what fits into the given scheme.

Or the moderator is not consistent in viewing and structuring. If no logical end is found after a brainstorming session, possibly with further open tasks, the participants lose the desire for this very efficient method.

Especially in abstract projects, such as development projects or scientific works, it is difficult to find a brilliant idea or the corresponding topic. Then, with the help of free association and brainstorming, you can send your thoughts on the road and approach the topic with a mind-mapping.

9.1.2 Mind-Mapping

Through a combination of figures, graphic elements (libraries, boxes) and keywords, the mind-mapping method ensures that the connections between the different aspects of the collection of ideas are clearly visible to everyone and their natural development remains comprehensible, so that the full overview of the implementation of the thoughts is always maintained.

A mind-mapping branches from the inside outwards and enables to structure work in neuronal structures corresponding to the right (creative) hemisphere.

The theory behind concept of mind-mapping supports the combination of the intuitive functioning of the right hemisphere with the structured functioning of the left. In this way, creativity can flow unhindered and at the same time it is possible to channel this energy, structure new thoughts and gradually transform them into concrete plans for measures. A mind-mapping is created in figure 9.1 for the functions of a control device.

It is therefore a tool with which new ideas can be recorded, organized and implemented.

Because planning does not only mean the determination of dates and deadlines. Rather, planning requires the ability to rethink a process as a whole in order to recognise the interplay of individual thoughts and to set priorities in such a way that only secondary activities are separated (chapter 6.5).

The branches of the mind-mapping can be accompanied by notes that further explain the text. Some programs has a direct link to MS Powerpoint®, MS Word® and MS Project®. This allows you to create a slide presentation from the mind map, a structure in MS Word or a project plan.

In this way, quite astonishing contexts can arise because an outsider, in this case the program, structures the examination.

The same effect is achieved when a newly written text is read out by the computer.

These methods are used in many areas - in the analysis and evaluation of organizational and management strategies as well as in finding ways to solve leadership tasks or in training in corporate and employee management.

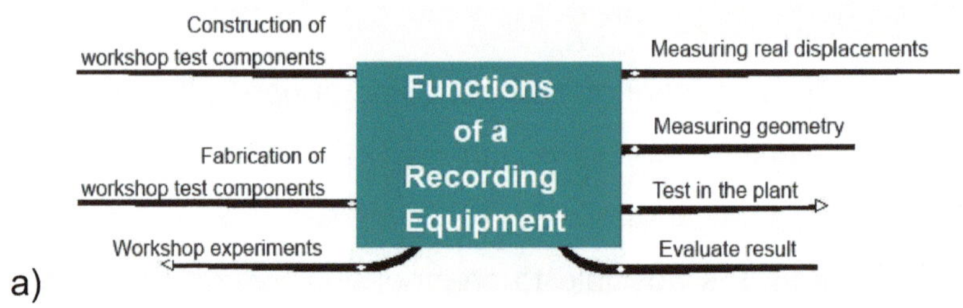

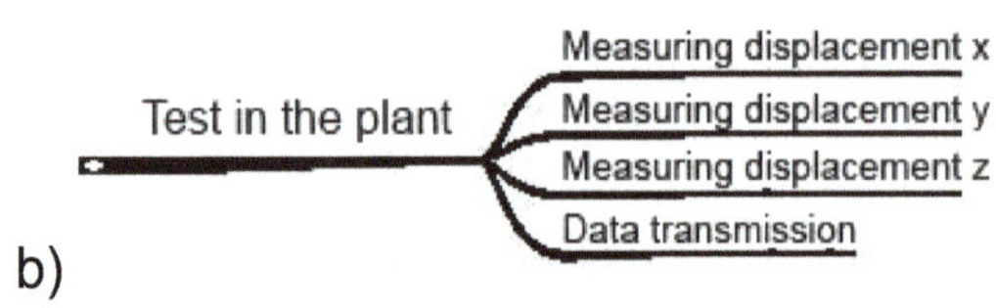

Figure 9.1: Mind-Map on the Functions of a Recording Equipment; a) Overview without Sublevels; b) Sublevel of the Main Branch „Test in the Plant"

9.1.3 MOSES Methode

The presenter Carmen Thomas /9.1/ hosted radio broadcasts on saturdays for decades. In order to come up with exciting ideas, she has developed the MOSES method. The abbreviation stands for man, place, concept, event and sensuality.

Now it becomes a term, in table 9.1.1.1 it is the abstract term "equilibrium", all that comes to mind spontaneously. From this a new story can be developed.

Table 9.1.1 MOSES for Balance

Human Being	Place	Concept	Event	Sensuality
Exclusions Broker Rests in it-self Does not move Just-divided Weighty Justice Scales Accord Dalai Lama	Seesaw Starving to death on the top of the seesaw Unequal-weighted Laotian Level Paradise	Tranquility Standstill Equal weight Equality Equation Pole Rope tensi-oning Stop	Stalemate Chess Settlement Liquidation Calm before the storm Equalization Stubs Overturning Handling	Gather-role together Crab Animal Plants Health Equal weight in nature Upsetting of the sea

9.1.4 The LEODARDO Principle

Leonardo da Vinci (1452-1519) is considered a universal genius of renaissance, the turn of the Middle Ages to modern times. He was a painter, sculptor, scientist and bold inventor. He has exhausted the full potential of his creative, technical-scientific and craft skills as few people understand it.

This makes it ideally suited as a role model for creativity and a wealth of ideas.

Micheal J. Gelb /9.3/ has translated the 7 principles of success, according to which Leonardo da Vinci lived, into a learning program. If you implement these 7 steps, you are much closer to your professional and personal success, because the unlived talents, which every person possesses, is activated again with this program.

Table 9.1.2 Die 7 Erfolgsprinzipien

Curiosià	**Dimostrazione**
The unquenchable curiosity for life and the constant search for new knowledge.	The willingness to test knowledge through experi-ence and perseverance and to learn from mistakes.
Sensazione	Sfumato
The constant sensitization of the senses, especially sight, as a means of reviving the experience.	The willingness to accept am-biguities, paradoxes and un-certainties. literally: go up in smoke

Continuation Table 9.1.2

Arte/Scienza	**Corporalità**
The development of the balance between science, art, logic and fantasy. Holistic thinking.	The cultivation of grace, am-bidextrousness, fitness and posture.
Connessione	
The realization that all things and	

phenomena are connected with each other. systemic thinking.	

Figures 9.2 and 9.3 show some of Leonardo da Vinci's famous works /9.3/.

Figure 9.2 Leonardo da Vinci; a) The Last Supper, 1495-1498; b) The Vitruvian Man, Proportional Study according to Vitruv, 1492; /9.4/

Figure 9.3 Leonardo da Vinci; a) Mona Lisa, 1503-06; b) Automobil /9.4/

9.1.5 ABC Method According to BIRKENBIHL

The ABC lists according to BIRKENBIHL can be used for stress-free repetition in class and for examination preparation.

First, two people write their knowledge into their individual ABC list (figure 9.4). This is a table whose 1st column is vertically numbered with the alphabet. The 2nd column remains for writing the information.

Table 9.5 ABC-List

A	
B	
C	
...	

Now the partners compare and complete their individual lists with each other's information.

This information is spoken, evaluated and further supplemented, perhaps even exchanged.

Furthermore, you can mark already understood parts green and unclear parts red to work on examination material.

Through the playful element in this method, people enjoy doing.

In my technical-mechanical lecture it is important to familiarize the students with the subject as early as possible and to head for the exams in exchange for learning. For this purpose, internships are introduced and certified.

The students were always very listless in calculating given taskness. And so was the result.

Since I only gave them topic groups and encouraged them to create their own tasks, they are full of zeal and calculate very complex tasks. It is their own interests that they dealt with.

9.2 Communication

Wherever people are closely related to each other, there is a large number of trouble spots that - unresolved - block part of our available energy.

People talk a lot about how they interact with each other, but there is much more non-verbal transport. These are on the one hand the chosen words, but also the tone of voice, the emphasis, etc. On the other hand, since only one-seventh of the communication takes place via the spoken word, most of the communication takes place via non-verbal contents, i. e. body language, gestures and facial expressions.

Professional communication makes it possible to perceive the behaviour and signals of the communication partners correct in order to be able to react appropriately and to determine the specific cause of his statement.

Schulz von Thun's Four-Ear Model provides an insight into how our statements are received by the person opposite us. The four sides of a message always apply: facts, self-disclosure, relationship and appeal, both to the sender and receiver.

In figure 9.4, we can really hear the little girl's indignation, although there is no acoustic reproduction. The body language alone, the posture, the eager arm and the facial expression, etc. give us information about it.

Figure 9.4 Nonverbal Communication /9.2/

Only the exact meaning of a statement is brought to the surface by the precise questioning behind it. For example, a targeted questioning of the respective interlocutors helps to find out the actual causes of a statement. It is very important to observe each other closely, because body language reveals the actual causes.

New communication techniques are constantly being tested, for example the ever more up-to-date tool mediation. By quoting the individual parties, a solution is worked out that does justice to everyone: the win-win principle.

In figure 9.5, the Four-Ear model of Schulz von Thun /9.5/, the contents of a message from the sender or receiver. Each message has different content aspects (figure 9.6). Depending on our experiences with the other person, we tend to react to messages in a certain pattern. Conflicts are therefore often preprogrammed (figure 9.7).

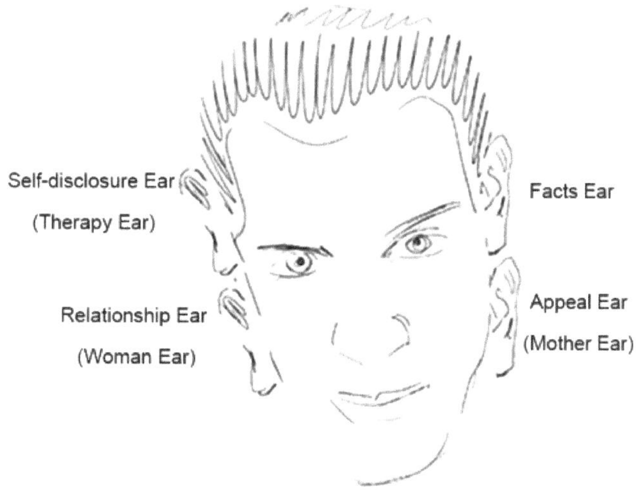

Figure 9.5 Four-Ear Model; Both as Sender and Receiver

If you are often confronted with such situations, your own communication skills must be trained. There are a number of methods for this purpose, which can be found in the literature under key qualifications. Here you can only give a brief insight into the topic.

The observance of Watzlawick´s three basic laws of communication is helpful here /9.6/

Figure 9.6 Four-Ear-Model by Schulz von Thun /9.5/; Sent and Received Message

- Communication does not know the opposite. We cannot not be non-communicating, every behaviour is interpreted as communication. There is no such thing as non-behavior.
- In case of doubt, the non-verbal message applies. If communication on the verbal and non-verbal levels is not congruent, the non-verbal statement is given greater weight than the verbal statement.

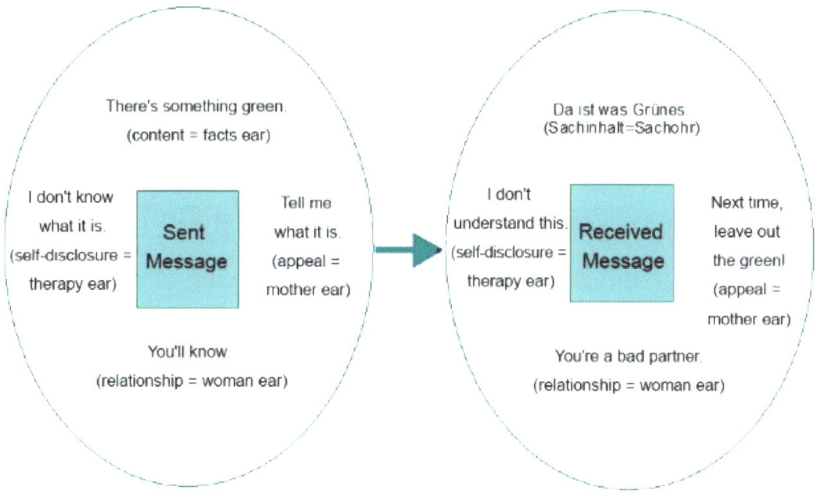

Figure 9.7 Four-Ear-Model by Schulz von Thun /9.3/; Contents of a Message from Sender to Receiver

- The received message is valid. Not what is sent, but what is received determines the further course of communication. The message sent is not available to the recipient.

Only according to these principles can communication succeed without conflict. The third rule in particular is immensely important for daily communication. Each person has his or her own "history" and should therefore be aware that his or her statements can be understood differently by the person opposite him or her.

For the project manager, it is important not only to find out the correcte project goal at the beginning of the project, but also to identify the actual cause of any disruptions during the course of the project in discussion with the team members.

A questioning technique based on the actual goal can help him. A definition of a goal can only be successful if the internal resistance of the participants is completely eliminated. In order to achieve this, it is necessary to "pick up the participants where they stand". This is best achieved when all five senses are addressed in the conversation.

Seeing Processes Holistically

Every person has his own preferred perception system. One of them likes to read long reports, but doesn't like to talk on the phone. His preferred organ of perception at work is the eye (visually) rather than the ear (auditive). His colleague just prefers the phone and hates long elaborations. If the two colleagues want to understand each other, they have to take this fact into account, otherwise they will "react" to each other.

9.2.1 Goal Definition

It´s easy to say "no!"
when there´s a deeper "yes!" burning inside
/6.2/

It applies

- Definition of objectives is essential.
- Without value definition no goal definition.
- What is important in my life?
- Question formulations for deepening:
    - What do you want to achieve?
    - What is your goal?
- How exactly do you want to be able to react to whom and when exactly?
- What do you want to know when you have reached your goal?

If we imagine that we are what we think, then it is not surprising that we ourselves have an effect on what falls to us (figure 9.8). So the same applies to our goals.

At the beginning of our life, we take over the system of values from our parents and their immediate environment. These are our beliefs, our guiding principles in life.

During puberty, it is the first time we react to ist and gradually develop our own value system.

In order to be able to develop ourselves further, we have to constantly questions back our value system behind.

Each of us has a brilliant flash of inspiration. But the chances and prerequisites for turning it into reality are not always equally distributed. A chance of fame and recognition, an idea for a new book, a new product or a marketing strategy, could not be exploited only because the correcte tool was not available to put the idea into practice beyond the critical planning phase.

Success means more than just having a great idea. Rather, it stands and falls with the ability to transform these thoughts into tangible goals, on the basis of which project plans can be worked out, milestones set and deadlines achieved. However, since goals can rarely be consistently pursued and achieved without taking the environment into account, flexibility and the ability to adapt to the constant change that a dynamic working environment is subject to is also required.

Figure 9.8 Beliefs

It is therefore necessary to define a precise goal. The method of Neurolinguistic Programming (NLP), a description, explanation and teaching model for communication, is process-oriented.

It is always possible to work without disclosing the goal. This is especially helpful when working in conflict situations in companies. Only the observation of his reactions and the background questioning with a certain questioning technique can show the experienced project manager how he can influence the behaviour of the employee without hurting his or her feelings.

In the "content-free" survey, the questioner does not know the target of the interviewee, as he or she does not want to or cannot disclose his or her target. However, the questioner must ensure that the interviewee has a sufficiently precise idea of the chosen target figure.

By questions like, "where exactly, how exactly, and to whom exactly?" a good analysis of the context is achieved, for example, at which point in time or to what extent the target behavior makes sense.

In order to get clarity for the goal, the desired state is described and addressed with all five senses (visual = seeing, auditive = hearing, kinesthetic = feeling, olfactory = smelling, gustatory = tasting) (figure 9.9) if possible.

For example, a car salesman sells a car with the following words: "Look at the great shape of the bodywork (V). And then the sound of the engine (A). And don't you sit very comfortably in the sports seats (K)? And smell the new leather (O)!" Only you can't taste the car (G)!

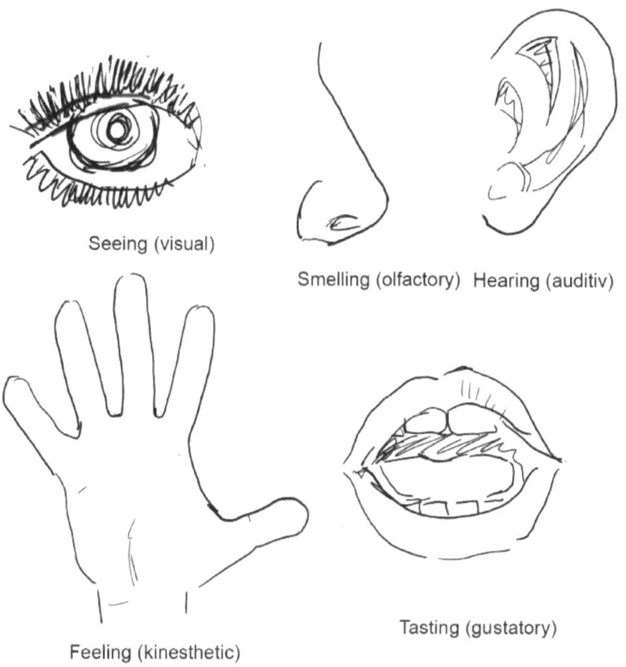

Figure 9.9: The Five Sensory Organs: Seeing, Hearing, Feeling, Smelling, Tasting

However, if the content of a goal is published, the questioner does not need to understand everything literally, because words are only metaphors whose real meaning is often not yet completely clear even to the goal seeker at the beginning of the process.

In addition, it is sometimes the aim of a discussion to initiate a search process to clarify or translate these metaphors.

If the goal changes during the discussion, this is usually accompanied by an increase in employee satisfaction. This becomes visible in his body expression, for example a symmetrical posture and a satisfied facial expression. The discussion leader encourages this process by helping the employee to formulate his or her goal more precisely. In this way, the statements become more coherent and the goal gains in contours. The employee looks reconciled and symmetrical. Quite different from the little girl in figure 9.4, he is now aware of his abilities.

Case Study

A group of students asked questions about excercises and I explained the solution. A study asked again and again, he obviously still did not understand the solution. So the process was repeated. When I finally asked whether all his questions had been answered sufficiently, he answered with "Yes", while his knee bounced up and down very quickly. "You may be, but not your knee not yet."

He confirmed it to me. He was embarrassed at the time to admit that he still didn't understand it. So we were able to approach the topic again in a relaxed way at another time.

9.2.2 Methods of Interviewing

9.2.2.1 Well-formed Goal Definitions

In problem situations, finding out the cause is usually an essential step towards solving the problem. It is a question of helping the interlocutor to move from a problem-oriented "way from" to a solution-oriented "way to".

With the help of a questioning technique, the desired goal is changed until it corresponds to a well-formed goal definition.

If there are changes in the goal during this process, it is desirable to do so, because the process of formulating the goal takes so long until it is coherent for the seeker and can be carried out with restriction in some or for all situations.

Then the abilities that would be particularly advantageous for this target state can be activated and a bridge can be built into the future.

9.2.2.2 Questioning Techniques for Goal Survey (Pene-Trance Model)

The Pene-Trance model in the NLP makes it possible to define a possible goal more clearly by means of specific questions. Since it can be used not only for the definition of a project goal, but also for conflicts within a project team, it is generally discussed here.

Questions for well-formed goal definition

The following questions help to describe the goal more clearly and define it better:

- What do you want to achieve?
- What is your goal?
- How exactly do you want to be able to react to whom and when exactly?
- What do you want to know when you have reached your goal?

The opponent is prompted to put himself in the situation in which he has already reached the goal. From this positive position, often unclear boundary conditions can be better understood and clarified.

Two examples are processed in parallel. The background of the first example "Objective of the development project introduction of a PPS program" (PPS) is that the project manager feels temporarily uninformed about the commercial facts of the project. He wants more control in the future, whenever he needs it.

The example is processed in table 9.2, table 9.4, table 9.6, table 9.8, table 9.10, table 9.12 and table 9.14.

This goal search can be discussed openly within the team.

The background of the second example "Self-initiated behavioral change of an employee" (SBE) is that a team member temporarily feels too personally attracted to the actions and statements of others, especially Mr. Müller.

This example is processed in table 9.3, table 9.5, table 9.7, table 9.9, table 9.11, table 9.13, table 9.15 to table 9.19.

He wants to have more control over his own reaction in the future, whenever he needs it. Since this search for a target is very personal and could also address personal weaknesses of him, the implementation is better suited for a private conversation.

Positive formulations (no negation)

Negative formulations are unsuitable for putting the seeker in a positive, desired goal situation. This only achieves positive formulations. The following possible questions help to formulate negative formulations in a positive way.

- How exactly will it be when you are/have no more "x"? (unstressed, no more being afraid, etc.)
- What are you have/feel instead?

It is important that the seeker has to find his or her own formulation and does not receive it from the moderator.

Table 9.2 Example; Objctive of the Development Project "Introduction of a PPS Program" (PPS)

What do you want to achieve?	I want to be in control of my development project at all times, for example, the current status of the budget.

What is your goal?	If I want to review the progress of the project in the evening, all relevant figures should be up to date so that I can sleep soundly.
How exactly do you want to be able to react exactly to whom and when?	With regard to my commercial project manager, I would like to be over delays in the figure at all times.
What do you want to know when you have reached your goal?	I want to have all relevant project data ready at the push of a button.

Table 9.3 Example: Self-Initiated Behavioral Change of an Employee (SBE)

What do you want to achieve?	I want to be able to listen to the meetings without getting upset when someone has made a mistake.
What is your goal?	To stay calm.
How exactly do you want to be able to react exactly to whom and when?	When Mr Müller talks about one thing again, I would like to be able to meet him calmly and objectively.
What do you want to know when you have reached your goal?	I would like to be able to concentrate on the content of the meeting and remain objective.

Table 9.4 Example; Objctive of the Development Project "Introduction of a PPS Program" (PPS)

How exactly will it be when you are no longer worried about the situation of the project?	I am in control at all times and feel like an actor.
What do you have instead?	I can edit the next tasks stress-free.

No Comparisons

Comparisons are also unsuitable. The following are possible questions for comparisons. They can be formulated by the following questions

- What do you know/can you know if you are "x"?
- What/how are you like if you are "x"?
- How will you know that you are "x" (safe, stress-free,...)?

Table 9.5 Example: Self-Initiated Behavioral Change of an Employee (SBE)

How exactly will it be when you are no longer impatient with your colleague?	I sit relaxed in the meeting and take objectively relevant notes.
What do you have instead?	I keep track of everything and can concentrate on the essentials, for example, the next actions to be carried out for the project.

Specifying the Behavioral Context

It is also always important to clarify at the end when and how the desired goal can be used.

- How, when, where and to whom should it be directed?
- When you have reached your goal, where and when, in which situation will you then act against whom and how?

But it can also be important to clarify who you don't want to behave towards!

Table 9.6 Example; Objctive of the Development Project "Introduction of a PPS Program" (PPS)

How do you know you're quiet?	I can focus on the things that lie ahead.
How are you when you're stress-free?	I'm lax, creative.
How will they know you're stress-free?	I whistle during work or I can respond well to others.

Table 9.7 Example: Self-Initiated Behavioral Change of an Employee(SBE)

How do you know you're relaxed?	I lean against the back of the chair and concentrate on what I have said.
How are you when you are relaxed?	I am friendly and have good ideas.
How will they know you're stress-free?	I am creative and find unconventional solutions.

Concretise Sense-Specific

It is necessary to find out how the seeker could feel when the desired goal is reached? This is an important point to achieve peace, especially in inter-human conflicts.

- How will you know when the goal is reached?
- What exactly will you see, hear, feel, smell or taste when you are about to achieve your goal? (V A K O G (figure 9.8)
- Where exactly do you feel it (in your body)?

Short Feedback Summary

The feedback on success should be brief, so that it is present in the memory and can be called up at any time.

- How can you see that your goal has been achieved?

Goal Experience, Goal Behaviour

In order to ensure that goal experience and behaviour can be initiated and maintained by the seeker time and again, the following clarifying question can help to define and fix the goal even more precisely:

- Assuming you could learn something that would bring you closer to your goal, what would that be?

Table 9.8 Example; Objctive of the Development Project "Introduction of a PPS Program" (PPS)

How, when, where and to whom should it be directed?	From my PC at my desk I can question the current commercial figures at any time.
When you have reached your goal, where and when, in which situation will you then behave as to whom?	If I go into the program and the current figures are available to me immediately, I can be happy about the progress of the project, even if everything doesn't run one hundred percent according to plan.

Questions about the Deepening and Strengthen of the Goal

This step is particularly suitable for working in one-to-one conversations, not so much within a group, because here "work on oneself" is being done, which may not be shown to the public.

In a one-to-one interview, for example with an unsatisfied employee, the other person can open up more easily and express his or her concerns and fears.

Every human being has abilities that can help him to reach or find a goal. He can recall them from his own experiences back into consciousness or he can observe them with other people and wish to possess them.

Table 9.9 Example: Self-Initiated Behavioral Change of an Employee (SBE)

How, when, where and to whom should it be directed?	When I realize that I am starting to get upset, and then I can decide for myself to curb my temperament in front of my colleagues.
When you have reached your goal, where and when, in which situation will you then behave as to whom?	I am also friendly and patient towards the colleagues who are getting me out of my mind today.

Table 9.10 Example; Objctive of the Development Project "Introduction of a PPS Program" (PPS)

How will you know that the goal has been achieved?	My questions regarding the current status of the project will be an-swered quickly and will stand up to the test.
What exactly will you see, hear, feel, smell or taste when you are about to achieve your goal? (V A K O G)	I am relaxed in the shoulder area (K). The teeth are loosely aligned (K). I am concentrated and quickly recognize the connections (V).
Where exactly do you feel it?	I am relaxed and friendly with my colleagues.

Table 9.11 Example: Self-Initiated Behavioral Change of an Employee (SBE)

How will you know that the goal has been achieved?	I perceive my anger and can con-trol it myself.
What exactly will you see, hear, feel, smell or taste when you are about to achieve your goal? (V A K O G)	I also see the abilities of my col-league (V), who otherwise gets me out of my mind, and I can stay friendly and patient with him (K).
Where exactly do you feel it?	I feel up to the situation. That makes me feel good, otherwise I have a stomachache.

Table 9.12 Example; Objctive of the Development Project "Introduction of a PPS Program" (PPS)

How can you see that your goal has been achieved?	My questions according to the current status of the project are answered quickly and withstand the test.

Table 9.13 Example: Self-Initiated Behavioral Change of an Employee (SBE)

How can you see that your goal has been achieved?	I am calm and relaxed and have a good feeling in my stomach.

Table 9.14 Example; Objctive of the Development Project "Introduction of a PPS Program" (PPS)

Suppose you could learn something to help you get closer to that goal, what would that be?	I could learn how to use the program.

Table 9.15 Example: Self-Initiated Behavioral Change of an Employee (SBE)

Suppose you could learn something to help you get closer to that goal, what would that be?	I learn to keep an eye on each other's abilities.

Development of Existing Skills

By means of specific questions such abilities can be made aware of themselves or a role model:

- What are your general abilities to get from a difficult condition to a good one? (at least three possibilities)
- What do other people do, what would other people do to achieve this?

Every behaviour has meaning in a certain context. It only becomes negative when the context no longer matches. This can be illustrated, for example, by the behaviour "screaming". In order to cry out for help, it is certainly a sensible behaviour. Shouting at one's children or employees, however, does not necessarily make sense.

Since every negative behaviour made sense at some point in time, this old behaviour has to be appreciated in this context. This gives the target seeker a choice and allows him/her to decide freely about his/her future behaviour.

Table 9.16 Example: Self-Initiated Behavioral Change of an Employee (SBE)

What skills do you generally have to get from a difficult condition to a good one?	I relax by sitting down loosely.
What do other people do, what would other people do to achieve this?	I'm taking a deep breath.

- As an experienced person/professional you have already experienced many things in your life, which can now be very useful in whatever form.
- When and where do you want to be able to experience the current behaviour unreservedly? (at least three possibilities)
- What is the good purpose of the old behaviour?

- What does it protect you from?

Table 9.17 Example: Self-Initiated Behavioral Change of an Employee (SBE)

As an experienced person, you have already experienced many things in your life that can now be very useful in whatever form.	I smile into myself and think of something positive.
When and where do you want to be able to experience the current behaviour?	When a colleague speaks out.
What good purpose does the old behaviour serve?	When danger is in delay.
What does it protect you from?	When colleagues appear unpre-pared in meetings.

This will also enable him to use his behaviour responsibly in the fu-ture.

- What consequences do you accept?
- Are you ready to pay the necessary price? Is it worth the ef-fort?
- Do you now have what you really want?

He thus assumes responsibility for his actions.

- How exactly will you know when the new and old behavior is displayed?
- What do you need to see, hear, feel, maybe smell and taste to know that it's time to try out the new behavior?

Table 9.18 Example: Self-Initiated Behavioral Change of an Employee (SBE)

What consequences do you accept?	I have to control my temperament now and then.
Are you ready to pay the necessary price? Is it worth the effort?	Yes, because I learn important things from my colleagues and get to know their skills.
Do you have what you really want now?	Yeah.

- Name three situations where what you have just learned is useful for you?

9.2.2.3 Further Clarifying Questions

The language has a surface and a deep structure. The surface structure describes what is literally said. The deep structure is what is really meant, i. e. the complete sentence in terms of content.

Table 9.19 Example: Self-Initiated Behavioral Change of an Employee (SBE)

How exactly will you know when the new behavior is displayed and when the old behavior is displayed?	When my colleague explains important findings, I remain calm and patient.
What do you need to see, hear, feel, maybe smell and taste to know that it's time to try out the new behavior?	I am relaxed and confident.
Do you name three situations where what you have just learned is useful for you?	In the progress review, in which many colleagues present.

If one has doubts about a statement, i. e. if false formations are detected (table 9.20 and table 9.21), these should be questioned in order to make the depth structure visible and thus make what is actually said accessible.

Malformities of Language

Formulations make it clear how the individual stands for something. In order to escape the pitfalls of a hopeless discussion, some answers to false formations are given in statements.

Problem-Solving Oriented Questions

The correcte question position (table 9.22) allows you to deepen your statement. But the opponent can also be lured out of the reserve. In this way possible problems can be addressed and solved.

Table 9.20 Redemptions and Generalizations

Redemptions	
Extermination	I'm glad.
Settlement repayment	He's dumber.
Adverb repayment	Obviously, they don't like me.
Modal operators	I have to, it's necessary, it's impossible.......
GeneraliZations	
Missing reference index	We don't want to get any more upset.
Universal quantifiers	all, all, all, nobody, everyone
Symmetrical predicates	argue, talk, fight (this includes at least two)
Asymmetric predicates	Smiling, slapping, stroking (only one actor necessary)
X or y	I have to worry about other people, or they don't like me.
Complex generalization equivalence	My colleague never smiles; he doesn't like me.
Incompletely specified verbs	My boss is driving me crazy.
Nominalizations	This decision to realize, I'm depressed.

Table 9.21 Distortions and Semantic Deformities

Presumptions	If he's as slow as his father, it won't do any good.
Cause-effect	My colleague makes me aggressive.
Implicit Causative	I want another job, but I don't have time for applications.
Thought-reading	She won't accept me.
Lost Performative	It is false to decide over and over again.

9.3 Conflict Management, Methods of Systemic Coaching and Systemic Constellation

In conflict situations, finding out the cause is usually an essential step towards resolution. It is a question of helping the interlocutor to move from a problem-oriented "away from" to "solution-oriented". Accurate questioning helps to detect and defuse sources of conflict.

Table 9.22: Detailed Questions on Some Statements

Predication	**Clearing Question**
I have a problem.	With what?
I don't know what to say.	What is the problem?
Everybody knows you can't win.	What do you have something to say? Really for everyone?

Continuation Table 9.22

Running away is not good.	What's good for you?
He's better for me.	For what?
It's impossible to trust any-one.	Completely impossible? Why is it impossible? When is it impossible?
One should respect others.	Says who? Why should you consider others? Why should one respect others?
All people are mean to me.	Who are all people? How did they get to you? Everybody?
I resent your questioning.	Which questions exactly? Are there specific questions?
The decision to come here was difficult.	How was the decision-making process? Why was the decision difficult?
If Paul keeps rejecting me, I won't be visiting him.	Why do you think Paul rejects you? What does one have to do with the other?
I'm worried about my col-league's beliefs.	What are you worried about? Exactly what beliefs?
I'd love to go with you, but my sister's sick.	What's your sister got to do with it?

Continuation Table 9.22

My husband never thinks about me.	How do you know he never thinks about you? What if he thinks about you? Never ever?
My colleague is arrogant.	How does this affects your behaviour?

9.3.1 Mediation

Mediators rely on consensus instead of compromise.

This is the assumption that conflicts are mostly due to deficiencies in communication. The mediator sits as a neutral mediator between the disputing parties.

For example, Mr. Mayer asks his boss for a long overdue salary increase. The boss refuses. The trial's off. Mr. Mayer complains to his wife and goes into internal dismissal. Engaging a lawyer would seem like a declaration of war against the boss, and a lawsuit would be expensive and tedious.

Unlike in court, which deals with the events of the past, and finally a winner and a loser, mediation takes a look into the future. The goal is a win-win situation, a solution that brings both sides a win-win situation.

Assuming that the boss's refusal is backed by a ruptured order, Mr. Mayer's childcare costs tear a hole in his little daughter's wallet. A change in his working hours and a later renegotiation or the establishment of a company kindergarten would be solution approaches that the negotiating parties could consider.

Behind mediation lies a structured scientific procedure that can also be used in business. The best-known mediation last years was the

German Lothar Geissler's attempt to reach an agreement on the reconstruction of Stuttgart's main station, which led to massive controversy among the population in the apron.

The advantages of mediation prevail over prejudices: Managers and entrepreneurs are increasingly giving a chance to convenient conflict resolution. Throughout Germany, there are numerous model projects for setting environmental standards, in project management and in the community.

In the USA, one-to-one interviews with the mediator are the rule, whereas in Germany the entire mediation procedure takes place before a plenary session. One-to-one interviews are the exception. The interests presented are in the foreground

- What are the reasons for the conflict?
- Why does the boss refuse a salary increase?
- Is there a solution that satisfies all interests?

A mediator must have empathy and sensitivity, whereas communication strategies, conflict analysis and the development of structural solutions can be trained.

People who make complex decisions in their professional everyday life, develop projects and design procedures or teamwork also benefit personally from the meditative structures.

A mediator does not always have to have professional competence in order to solve a conflict, because they do not make any professional decisions, but accompany the solution, for example by consulting an expert.

A mediator works on the conflict structure. Disputes between two companies have a different structure than internal conflicts, in which personal strife and positions of power often play a role.

This is where systemic coaching and systemic positioning come into play.

9.3.2 Methods of Systemic Coaching and Systemic Constellation

These personal entanglements are "handed over" to a person in different ways.

Every human being lives in reference systems such as family, faith or culture, in which he is confronted already in his earliest childhood, or like a company or circle of friends that he chooses himself/herself. There are profit systems like companies or non-profit systems like the family.

These systems influence one's own life through the system laws, for example knocking at the door before entering the other's territory. If these laws are violated, conflicts arise.

This applies to both the private and the company.

It can even be said that what is not solved privately in entanglement is carried into the company in order to be active there. This makes it very easy to explain many power struggles.

These entanglements are made visible by a systemic constellation.

(System) Installation Types

A wide variety of (system) installation types are available for this purpose

- The big constellation with real people as representatives,
- small installation inside with one person,
- with figures or objects, (the line of vision must be clear),
- with anchors, e. g. chairs to put in or floor plates for placing on top of it,
- virtually in the head (visualize with closed eyes inside),
- large and small analysis,
- concealed (representative does not know who it is) or open,

- medication example (all medications with numbers),
- with and without analysis.

The representative shall be a person or object involved in such involvement.

Constellation Permit

In order to make a constellation, a constellation permission must be given and it must be clarified, what the actual topic is, which is worked on.

It is to be distinguished whether someone is doing something or not,

- because he cannot do it,
- because he does not want it or
- because he is not allowed to do it.

“I can't" is about the training area. The person can practice, learn and train, of course with role-playing or other methods. Here is the sequence learning leads to knowledge, which in turn leads to action.

"I don't want to" on the other hand, shows that it doesn't make sense to the person, doesn't benefit you. Here the attitude towards negative attitudes can be changed at coaching level.

The lowest level is "I not able to (I can´t)". Here the system does not allow something. For example,in most of the countries you don't drive over a red traffic light, in some other countries no one would ever wait patiently for 5 minutes at night at a red traffic light while the road is empty and not a human being to see.

In the same way, many other unconscious processes that are shaped in the family home are lived. These are then also lived in companies with colleagues.

If someone is not allowed to do something, it's time to get permission. This is systemic coaching.

Case Study

A young man can get a permanent position at least for 6 years, if he has successfully completed his dissertation procedure by a certain date. His doctoral thesis was almost completed two years ago. He pretends to have no time when his doctoral supervisor suggests that he should take a week's leave for the completion. Technically, he is brilliant and doesn't have to shy away from the doctoral examination.

A working hypothesis could be that the system does not allow him to move into another social class. Maybe his family rejected "students", maybe his father had not made it to the top? These theses must be examined.

System Laws

Different system laws can be affected in the company organization.

- In projects, the principle described above must be clarified: a person cannot, does not want or is not allowed to do so.
- Often there is arrogance:"I can do better".
- Failure to comply with the ranking order results in harass-ment.

Family conflicts are often reenacted in the context of work. To find out (analysis work), you look behind the scenes. It's about bringing things to light that you never thought of.

Then a working hypothesis is created and asked for.

- What is that good for?
- What does it stand for?

This working hypothesis may change during the constellation process due to new aspects.

The best thing would be to systematically set up the project team at the beginning of a project in order to improve project management. With questions like

- Who doesn't want to?
- Who could play a role that I hadn't thought of?
- What is it really about?

Unclear project areas are illuminated. After the analysis further questions will be asked

- Which game is in progress?
- How can a system be used? For example, as an army, not as an open rebellion.
- How can the system be operated?

Case Study

A student postpones his last exams again and again. There are reasons, how a serious illness at the examination date and at the next date the favorite dog dies the day before.

Good reasons. And yet I'm still on my way, because I suspect that something is hidden behind it. "You don't have to answer me, but please think about it. Can you be afraid to get ready?"

Silence.

"You ask me a question!"

And he starts talking about what's bothering him. During an examination he was publicly accused of being never a good engineer by a lecturer. The exam grade was a good one.

The systemic laws work subconsciously, in secret. They are not usually conscious to humans. Nevertheless, their effect can be felt by everyone and/or the whole system in a positive or negative way. If these system laws are adhered to, the whole system is motivated.

The relationships are voiced, each individual feels supported, strengthened, etc. If they are ignored, the system and each individual are weakened.

- The most important system law is "belonging to one's own system", because that means survival. This is still the case in the animal kingdom today, and we bring this heritage with us, which is stored in our behavioural programmes.
- The second most important system law is "love and recognition, ". No system can function in the long term without recognition.

System Laws According to Hanisch

The principle is first the order, then the binding and finally the appreciation. Here, those that take place within the company are explained in more detail.

- Involvements
- Compensation
- Ranking order
- Who goes, who stays?
- Belonging
- Binding
- Love and recognition
- Identification
- Double exposure
- Succession
- Arrogance

9.3.1 Involvements

For a system to work, you have to have an order. In the business context, it is the

- Order of recognition
- Order of Power

An invisible order leads to entanglement.

9.3.2 Compensation

If you don't get enough money for a job, this leads to

- Internal termination,
- regulatory service,
- work according to regulations (take your time) and
- theft (one takes what one deserves).

If you get too much money for a job, that leads to

- Arrogance and
- others feel disadvantaged and/or treated unfairly.

In sales and distribution, it is always a question of giving the customer something more to bind him/her. In marriage, housework must be balanced.

The family turns away when a family member to whom you have borrowed money cannot repay it.

9.3.3 Ranking Order

Outside the family, the ranking order, for example the hierarchy in the company, is also regulated by system laws.

If you sit on the executive´s chair, that means pushing forward. There is the right of the elder or the right of the first.

A new collegue is the last in the system. It is therefore a systemic mistake to mark the collegue at the beginning. As a result, the desire of our employees may grow "we saw him off". In this case, the system has to lead from behind, which means that the first place has to be earned by him.

9.3.4 Who Goes, Who Stays?

In a company, the person who no longer makes or is unable to make a contribution must leave.

In the case of quitting, the rule is that whoever came last must go first.

The company has to let someone else be dignified, or the system will avenge itself. If someone is not dismissed with dignity, those who are left will find it disgraceful. This leads to revenge in the company.

The company's goal is to stay alive economically.

9.3.5 Belonging

It is defined by language, culture or religion. The integration test is an example of testing whether there is enough information on the country of immigration. Surprisingly, many locals couldn't answer the questions at all. But it is not about the questions, it is about systemic acceptance of national identity.

For example, the belonging is determined by murder.

In the company, this means that damage is deliberately done to the company.

If someone has not accepted the rules of a system, he is excluded, for example, formerly men who went on paternity leave. It was only gradually that the opinion changed. However, it is not yet fully accepted. Very few men go on parental leave.

Fusions usually don't work, as we have seen again and again in recent years.

An integration process would have to take place by appreciating the old, the old logo, etc., for example in a farewell speech (for example, with a ritual). Otherwise, the employees lose their membership and then their right to belong.

The company has to give something first, then the employees give.

Rituals hold life together. They are used to organize living together.

9.3.5 Binding

If you are separated from a loved one by whatever, but still love that person, this connection must be solved by a farewell ritual with the person, for example marriage with the ritual of divorce. Binding can also be to a dead body. Hate is also a bond.

However, there are also employees who are bound to the company. You can't leave before the bond is broken.

The evaluation of the solution is particularly important here.

9.3.7 Love and Recognition

If love and recognition do not flow, human and his soul cry out for it. If it continues to be absent, human goes into the disease.

Unwanted children are denied the right to belong, they look for replacements elsewhere and can go into addiction. Addiction is the search for something.

The question,"What did you get from your father and mother?" is asked. This will be care, but also often money as compensation for giving and taking.

All this information (thoughts) is stored energetically and the subconscious knows it.

In the company, recognition of performance stands for love. The consequences are comparable.

Case Study

Why do millions of people get involved with Facebook and invest a lot of time and effort? The answer is visibility, reward and feedback that comes quickly and promptly.

Visibility, the visualization of the ego brand, is an enormous driver at work. The personal strengths (each person is unique) are thus made visible.

In doing so, the reward is an important factor. And it is not just about money or payment, but much more about appreciation and recognition of the work done. You get used to higher pay fast. That's why managers and colleagues should praise and appreciate when their work performance was good.

Praise and appreciation will be welcomed as positive feedback. Feedback is always also the comparison with others and serves as an incentive, as a learning aid by role model. This is necessary for personal development.

9.3.8 Identification

If someone has been unjustifiably expelled in the system, it can happen that someone else becomes like that, unconsciously or consciously. On an unconscious level, this is done.

Identification is taken over if a person has wrongly lost the right to belonging, for example migrant children.

9.3.9 Double Exposure

Someone sees someone else in you. For example, if there is a rebellion conflict with father or mother, there may be another rebellion with the superior, for example in the wife or husband or chief.

9.3.10 Succession

If someone follows and the new one follows, the recognition of the predecessor must take place, otherwise the system will be beheaded.

The question applies: Who has the right to succeed? Who's moving up?

The old takes precedence over the new.

Here the old must be appreciated and looked at, which is still useful. This is the only way to develop.

This is important for consultants in companies. Without appreciation of the old working methods, reforms are boycotted, especially if they were developed without the cooperation of the staff.

9.3.11 Arrogance

It's like Goethe's sorcerer's apprentice.

The apprentice thinks he's already better than the master. In a company context, the apprentice can also be a pupil, student or employee, the master can also be a teacher, professor or supervisor.

But also given advices give without order, are presumption.

"Advice is also beatings."

It means,"I can do it /is better than you."

Accepting and then believing to be able to do it better is the pre-stage for megalomania.

9.4 Test Your Progress

Here you have the opportunity to check your knowledge. In table 9.23 you can answer the questions.

Table 9.23 Test for chapter 9: What statements are correct or false?

No.	Question	cor-rect	false
1	Brainstorming is about quality over quantity.	☐	☐
2	A mind map allows unstructured work with neuronal structures corresponding to the right hemisphere of the brain.	☐	☐
3	The exact background questioning helps to detect and solve these sources of conflict.	☐	☐
4	The aim of a discussion is often to initiate a search process to clarify or translate a target.	☐	☐
5	The human being has six senses.	☐	☐
6	The pene-trance model can only be used to define a project target.	☐	☐
7	The person seeking advice or the team member does not have to take responsibility for his actions.	☐	☐
8	Repayments and generalizations must be questioned in order to obtain the actual content of a statement.	☐	☐
9	The ranking can be changed quietly.	☐	☐
10	Whoever comes to the company first, goes first in times of crisis?	☐	☐
11	It's completely unproblematic if you get too	☐	☐

	much pay.		

1	**2**	**3**	**4**	**5**	**6**	**7**	**8**	**10**	**11**
f	r	r	r	f	f	f	r	f	f

10 Résumé

Projects are time-limited companies with a clearly defined beginning and end. This distinguishes them from the continuous work of an organization.

Each successful project is divided into four phases: definition, planning, execution and completion (figure 10.1).

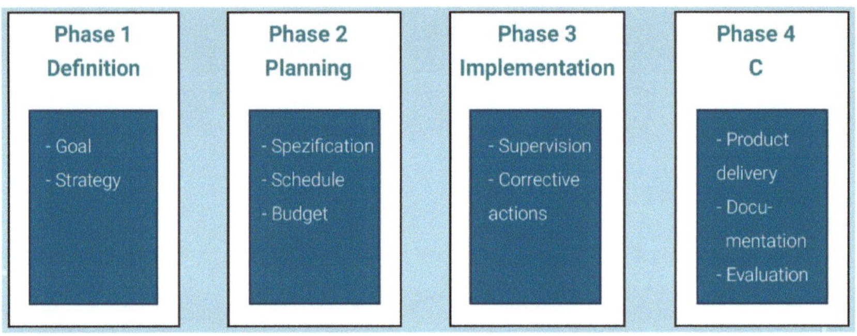

Figure 10.1: Four Phases of a Project

It is extremely important for the success of a project that it is clearly defined before it starts. Each definition includes the performance criteria that determine the successful completion of the project. Once the project is under way, changes will inevitably be made to these criteria. However, it is important that all changes with the foreseeable effects on schedule and budget are contractually recorded.

A successful project leads to the expected result at the expected time and within budget. For this reason, the three project parameters are defined in the planning phase: The specifications define the quality standard. The deadlines are set by an exact schedule, the necessary costs are fixed in a budget.

A project team is put together to carry out the project. This requires a project organization. The tasks and responsibilities are distributed. The employees are instructed in their tasks. In the case of large projects, procedures and guidelines (job procedures) are drawn up to ensure the team's activities during implementation.

The project manager has not only at the beginning of the project many different tasks (table 10.1). He must coordinate the work of the various groups and monitor the progress of the project. In the event of deviations from the planned values, he must take corrective action. He is expected to motivate the team by giving feedback, conducting the necessary negotiations and helping resolve disagreements.

The project goal is to deliver the agreed result to the client. Before the project is completed, however, there are a number of tasks to be completed: writing the documentation, providing operator training, disposing of surplus material and equipment, evaluating the experience gained, concluding with an acceptance test, writing a project completion report and carrying out a project review with the core of the project team.

But not every project demands the same effectiveness in this activity. It depends on the type, size, scope and organization of the project in which it is involved. Each project manager will decide for himself which measures make sense.

Table 10.1: Task List for a Project

1	Define the project.	☐
2	Choosing a strategy.	☐
3	Developing specifications.	☐
4	Develop a schedule.	☐

Continuation Table 10.1

5	Developing a budget.	☐

6	Organizing the project team.	☐
7	Assigning Taskn and duties.	☐
8	Training new employees.	☐
9	Monitor progress.	☐
10	Take corrective action.	☐
11	Provide feedback.	☐
12	Test final result.	☐
13	Deliver results to the client.	☐
14	Writing operating instructions.	☐
15	Training of operating personnel.	☐
16	Transfer project team members.	☐
17	Disposal of surplus equipment and material.	☐
18	Leave facilities and premises free.	☐
19	Evaluate project performance.	☐
20	Carrying out the final inspection.	☐
21	Completing the project report.	☐
22	Carrying out project review with the core employees.	☐

Literature

2.1; Doran, G. T.; There's a S.M.A.R.T. way to write management's goals and objectives. Management Review, Volume 70, Issue 11(AMA FORUM), pp. 35-36, 1981

2.2; http://www.psy.lmu.de/soz/studium/downloads_folien/ws_09_10/muf_09_10/muf_schattke_0910.pdf

2.3; Schelle, Heinz; Projekte zum Erfolg führen, dtv Verlagsgesellschaft; edition: 7, 2014

2.4; https://wissensarbeiter.wordpress.com/2015/06/30/produkte-und-projekte-was-sind-die-unterschiede/

4.1; Pantalon, M.V.; Nicht warten- Starten – Das 7-Minuten-Programm zur Motivation, dtv premium, 2012

5.1; Hofstetter, H.; Der Faktor Mensch im Projekt, In: Schelle, H., Reschke, H., Schnopp, R., Schub A. (Hrsg.): Loseblattsammlung "Projekte erfolgreich managen", Veröffentlichungen des Verbands Deutscher Maschinen-und Anlagenbau e. V., Köln, 1998

6.1; Seiwert, Lothar J.; Das 1 x 1 des Zeitmanagements, Knaur Ratgeber, 2014

6.2; Covey, Stephen R. /Merrill, A. Roger /Merrill, Rebecca R; First Things First, Fireside by Simon & Schuster, New York, 1995

6.3; Covey, Stephen R. /Merrill, A. Roger /Merrill, Rebecca R.; Der Weg zum Wesentlichen: Zeitmanagement der vierten Generation, Campus Verlag, Frankfurt /M., New York, 2014

6.4; Francis C.; Young D.; Tuckman & Jensen, 1977

6.5; Young; A Technique for Producing Ideas, Mcgraw-Hill Professional, 2003

6.6; Motamedi, Susanne; Konfliktmanagement vom Konfliktvermeider zum Konfliktmanager, Grundlagen, Techniken, Lösungswege, Offenbach, 1st edition, 1999

6.7; Wazlawick, Paul; Anleitung zum Unglücklichsein, Piper, 2009

6.8; Süß, Gerda; Eschlbeck, Dieter; Der Projektmanagement-Kompass, So steuern Sie Projekte kompetent und erfolgreich, Braunschweig, 1st edition, 2002

6.9; Birker, Gabriele; Birker, Klaus; von Pepels, Werner (Hrsg.); Teamentwicklung und Konfliktmanagement, Effizienzsteigerung durch Kooperation, Berlin, 1st edition, 2001

6.10; Haeske, Udo; Team-und Konfliktmanagement, Teams erfolgreich leiten, Konflikte konstruktiv lösen, Berlin, 2013

8.1; Kuppinger, Martin; Reister, Steffen; Jäger Matthias; Reister, Steffen; Microsoft Project 2002-Das Handbuch -, Microsoft Press Deutschland, 2003

8.2, Orthwein, M.; Meßmer D., Agiles Projektmanagement – Projektentwicklung mit Scrum, Kanban & Co., Techdivison, Internet Stand 2015-12-06; (https://www.techdivision.com/_Resources/Persistent/a90c984a454ba0b8478694b83f7a8822514b8fc8/Agiles-PM-Whitepaper0502.pdf)

8.3, Kriegisch, A., Scrum - Auf einer Seite erklärt, Internet Stand: 2015-12-01, (http://scrum-master.de/Was_ist_Scrum/Scrum_auf_einer_Seite_erklaert)

8.4, Kriegisch, A.: Scrum-Rollen–Product Owner, Scrum Master, Internet Stand: 2015-12-01, (http://scrum-master.de/Scrum-Rollen/Scrum-Rollen_Product_Owner)

8 . 5, Alphanodes: Meindl, C., Scrum-Rollen – Der Product Owner, Internet Stand: 2015-12-01, (https://alphanodes.com/de/product-owner)

8.6, Denkmotor: Was ist Design Thinking?, Internet Stand 2015-12-05, Video, (http://www.denkmotor.com/angebot/kreativitat-und-innovation/seminardesign-thinking/)

8.7, Bozyazi, E.: Design Thinking im Projektmanagement, Lehrmaterial Hochschule, Mannheim, Oktober 2014

9.1; Thomas, Carmen; ModerationsAkademie für Medien und Wirtschaft, Engelskirchen

9.2; ; Acer-Werbung, FAZ 2000

9.3; Gelb, M. J.; Das Leonardo-Prinzip, Econ Tb, 2001

9.4; ; Wikepedia

9.5; Schulz von Thun, Friedemann; Miteinander reden, rororo Rowohlt-Verlag, Reinbek bei Hamburg, 2010

9.6; Wazlawick, Paul; Beavin, Janet H.; Jackson, Don D.; Menschliche Kommunikation. Formen, Störungen, Paradoxien, Huber Hans, 2011

9.7; Schmidt-Tanger, Martina; Kreische, Jörn; NLP-Modelle-Fluff & Facts, VAK Verlag für angewandte Kinesiologie GmbH, Freiburg im Breisgau, 2005

Additional Literature

Arden, Paul; Egal, was du denkst, denk das Gegenteil, Bastei Lübbe (Lübbe Ehrenwirth); edition: 5 (2011)

Berckhan, Barbara; Die etwas andere Art, sich durchzusetzen-Selbstbehauptungstraining für Frauen; dtv, 2003

Birkenbihl, Vera; Trotzdem lehren (MVG Verlag bei Redline), Moderne Verlagsges. Mvg, 2013

Cameron, Julia; Von der Kunst des Schreibens und der spielerischen Freude, Worte fließen zu lassen, Droemer Knaur, (2003)

Csikszentmihalyi, Mihaly; Kreativität: Wie Sie das Unmögliche schaffen und Ihre Grenzen überwinden, Klett-Cotta, 2015

Dörner, Dietrich; Die Logik des Mißlingens-Strategisches Denken in komplexen Situationen, rororo, 2015

Dulabaum, Nina L.; Mediation: das ABC, die Kunst, in Konflikten erfolgreich zu vermitteln, Weinheim, 4th edition, 2009

Ernst, Heiko; Können wir unserem Bauchgefühl vertrauen?, Psychologie heute, 03/2003, S.20 ff

Eyer, Eckhard (Hrsg.); Report Wirtschaftsmediation, Krisen meistern durch professionelles Konflikt-Management, Düsseldorf, 2004

Glasl, Friedrich; Konfliktmanagement, ein Handbuch für Führungskräfte, Beraterinnen und Berater, Bern, 7th edition, 2002

Goldberg, Natalie; Schreiben in Cafés, Natalie Goldberg, Autorenhaus, Auflage: 2nd edition, 2009

Hanisch, Christian; Ausbildung zum Aufstellungsleiter „Systemischer Coach ICI /Systemischer Aufstellungsleiter ICI", European Business Ecademy, Seven Mirrors Consulting GmbH, 2011

Hansel, J.; /Lomnitz, G.; Projektleiter-Praxis-Erfolgreiche Projektabwicklung durch verbesserte Kommunikation und Kooperation, Springer-Verlag, Berlin, 1987

Haynes, Marion E; Projekt-Management-Von der Idee bis zur Umsetzung -, Wirtschaftsverlag Carl Ueberreuter, Wien, 2003

Kotter, John; Rathgeber, Holger; Stadler, Harald; Das Pinguin Prinzip-Wie Veränderung zum Erfolg führt-, Droemer, 2011

Küstenmacher, Werner Tiki; Simplify your life-einfacher und glücklicher Leben-; Knaur TB, 2011

Litke, Hans-D.; Projektmanagement: Methoden, Techniken, Verhaltensweisen, Carl Hanser Verlag, München /Wien, 2015

Lohmann, Friedrich; Konflikte lösen mit NLP, Techniken für Schlichungs-und Vermittlungsgespräche, Paarberatung und Mediation, nach Virginia Satir, John Grinder und Thies Stahl, Paderborn, 2003

Lundin e. a.; Fish-Ein ungewöhnliches Motivationsbuch-; Ueberreuther Wirtschaft; 2015

Massow, Martin; Gute Arbeit braucht Zeit-Entdeckung der kreativen Langsamkeit; Heyne, 1999

Mehrmann, Elisabeth; Wirtz, Thomas; Effizientes Projektmanagement, ECOn Taschenbuch Verlag, Düsseldorf, 2nd edition, 2002

Meise, Sylvie; Hör doch mal zu!, Psychologie heute, 07/2003, S.46 ff

Michel, Reiner M.; Projektcontrolling und Reporting, Sauer-Verlag, Heidelberg, 1989

Modler, Peter; Das Arroganz-Prinzip: So haben Frauen mehr Erfolg im Beruf, Krüger, Frankfurt, 2012

Motamedi, Susanne; Körpersprache – schwere Sprache, Psychologie heute, 10/1996, S.52 ff

Mühlisch, Sabine; Das Prinzip KörperSprache im Unternehmen: Inspirationen für eine lebendige Arbeitsgestaltung, Junfermann, 2014

Naumann, Frank; Diplomatie: Der sanfte Weg zum Sieg, Psychologie heute, 11/2003, S 64 ff

Nöllke, Matthias; Schlagfertigkeit; Haufe; 2015

Rahn-Huber, Ulla; Der Vampir neben dir; Kreuz, 2002

Schwarz, Gerhard; Konfliktmanagement, Konflikte erkennen, analysieren, lösen, Wiesbaden, 2013

Seiwert, Lothar J.; Wenn Du es eilig hast, gehe langsam, campus, 2012

Stoss, Karl; Botschen, Günther; Management der Strategischen Geschäftseinheiten, Verlag Gabler, Wiesbaden, 1994

Weidner, Christopher A.; Feng Shui gegen das Chaos auf dem Schreibtisch, rororo, 2004

Weyer, Simone; Konfliktmanagement im Projekt, Diplomarbeit FH Bochum, Fachbereich Wirtschaft, 2004

Will, Franz; Teamkonflikte erkennen und lösen: Zwischen Emotionen und Sachzwängen, Beltz; 2012

Wolf, Axel; Macht: Wer dominiert wen?, Psychologie heute, 01/1999, S 20 ff

FEEDBACK

Thank you for a positive review

If you liked the book, please send me a positive review on Amazon Kindle.

Comments, questions or criticism

Here you can send me your comments, questions or criticism about the book "7 Boyant Checklists for Your Project".

In the Google form you can write me directly and you can book a Strategy Session here.

Appendix: Forms

Click here for free bonus material on the book.

1 Project Organisation

Form 1.1: Project Completion Report

Project:		**Project Number:**
Project Manager:		**Date:**
Project Goal:		
Project Organization:		
Description of perfor-mance:		
Important events /critical problems:		
Project costs:		
Signatures:	Management	Project Manager

Form 1.2: Detailed Activities of the Poject

Project Manager:					**Project Number:**		
					Date:		
					Page:	Date	
	Ope-ration /Procedure	Start	End	Dura-tion	Running. No. of Pre-decessor	Resource	Re-quired Time
1							
2							
3							
4							
5							

Form 1.3: Project Proposal

Project:		**Project Number:**		
Project Description:		**Scheduled Start:**		**Scheduled End:**
Project Manager:		**Project Team:**		
Time and Effort Estimation		**Hours**	**€**	
	Labour			
	Overhead			
	Stock			
	Equipment			
	Equipment Rentals			
	Incidental expenses			
	Profit			
Total Costs				
Annexes				
Profit Analysis ☐	Calculation ☐	Capital Expenditure ☐	Risk Analysis ☐	

Continuation Form 1.3

Date:	Proposer:			

Approval:	Management:	Date:	

Form 1.4: Project Manager Tasks/Competencies for a Project

<table>
<tr><td>Project:</td><td colspan="3">Project Number:</td></tr>
<tr><td colspan="4">Project Description:</td></tr>
<tr><td>Project Manager</td><td colspan="2">Deputy:</td><td>Project Team:</td></tr>
<tr><td colspan="4">Total Project Costs:</td></tr>
<tr><td>Tasks</td><td colspan="2">Planning/Implementation of the project:</td><td></td></tr>
<tr><td></td><td colspan="2">Project organization / documentation</td><td></td></tr>
<tr><td></td><td colspan="2">Work delegation to the specialist departments</td><td></td></tr>
<tr><td></td><td colspan="2">Monitoring/control with regard to perfor-
mance/dates/costs</td><td></td></tr>
<tr><td></td><td colspan="2">Report to the steering committee</td><td></td></tr>
<tr><td>Competence:</td><td colspan="2">Entitled to sign at the level of the project budget</td><td></td></tr>
<tr><td></td><td colspan="2">Countersignature by the executive board from
€...
.......</td><td></td></tr>
</table>

Continuation Form 1.4

	Authority to issue instructions to the project team, including the project-related authority to issue instructions to supervisors integrated in the project	

	team	
Date:	Signatures:	

Form 1.5: Members of the Project Team

Project:				
Project Number:				
Date:				
Name	**Abbrevia-tion**	**Telephone**	**Mobile**	**E-Mail**

Form 1.6: Organization of the Project Team

Project:		**Project Number:**		
Project Manager:		Date		
		applies	don't know	does not apply
1	Have a project manager (and their deputy) been appointed?			
2	Does the project manager already have project experience?			
3	Have special project managers (and their deputies) been appoint-			

	ed?			
4	Does the project team already have project experience?			
5	Has the scheduling been coordinated with the project team?			
6	Is the schedule regularly revised and regenerated?			
7	Are project meetings held regularly?			
8	Has the project progress been coordinated with the project team?			
9	Has the project team been made aware of the obligation to provide information?			
10	Is the specified release and modification procedure adhered to?			
11	Are suppliers regularly checked?			

Form 1.7: Risk Analysis During Project Planning

Project Manager:			
Project:			
Project Number:			
A	Technical Risks	Material composition	
		Material availability	
		Machine utilization	
		Mounting	
B	Economic Risks	Financing	
		Contracts	
		Information	
		Documentary	
		Patents	
C	Political Risks	Which jurisdiction is applicable?	
		Which import/export regulations apply?	
		What customs restrictions apply?	

Continuation Form 1.7

		What currency is used for settlement?	
		Capital transfer?	
D	**Risk Factor "Human Being"**	Project manager	
		Project team	
		Expertise project manager/team	
		Know-how project manager - team	

Form 1.8: Feedback on Project Completion

Project:		**Project Number:**				
		Date:				
Decision Maker		**Control Circuit?** ☐				
		Grades(1=good; 5=worst)				
		1	2	3	4	5
1	How satisfied were you with the first speech, description of the task setting and the goal setting by the project team?					

Continuation Form 1.8

2	How satisfied were you with the support of your project management and the planning team?					
3	How satisfied were you with the communication in the project groups or coordination meetings?					
4	Were important functionaries or decision-makers not included in the project structure?					
5	How much do you value your contribution to problem solving in terms of defining goals?					
6	How do you assess the project with regard to the realisation of all partial/project objectives?					

Form 1.9: Monthly Report of the Project

Project Number:	**Date:**		**Responsible:**
Performance in the reporting month:			
Deadline situation:			
Deviations:			
Measures envisaged:			

Form 1.10: Requirement Profile of the Project Manager

Name:			

	Criteria	high	me-dium	low
1	Professional competence			
1.1	Technical know-how			
1.2	Knowledge of the corporate structure			
1.3	Knowledge of the information channels			
1.4	Knowledge of operational structures			
1.5	Project management experience			
2	Methodological expertise			
2.1	Knowledge of planning techniques			
2.2	Knowledge of the methods for capacity/cost planning			
2.3	Knowledge of milestone trend analysis			
2.4	Knowledge of risk analysis			
2.5	Knowledge of change management			
2.6	Knowledge of contract management methods			

Continuation Form 1.10

2.7	Knowledge of quality management			
2.8	Knowledge of presentation techniques			

2.9	Acceptance in the company			
2.10	Knowledge of creativity methods			
2.11	Knowledge of communication techniques			
2.12	Knowledge of self-management			
3	Social competence			
3.1	Initiative			
3.2	Determination			
3.3	Assertiveness			
3.4	Empathy			
3.5	Organizational talent			
3.6	Negotiation skills			
3.7	Motivational ability			
3.8	Delegation capacity			
3.9	Critical faculty			

Continuation Form 1.10

3.10	Conflict ability			
3.11	Willingness to learn			
3.12	ability to communicate			

3.13	Active listening			
3.14	Expressed comprehensibly			

Form 1.11: Risk Analysis During Project Implementation

Project:				**Project Number:**			
Project Manager:		**Signiture:**		**Date:**			
Task	Risk	Risk Level of risk 1-10 (low-high)	Cause	Effect on	Actions	pre-ven-tive	cor-rec-tive

Form 1.12: Job Description of a Project Manager

Job Description				
Last name, first name, date of birth			Function name	
			Status	
Company	Division	Group	Job No.	Tarif group

Description of the task:				
Employee reports to:				
The holder of the position receives specialist instructions from:				
Directly subordinate employees (surname, first name, function, introduction):				
The holder of the position indicates his or her professional instructions (name, function):				
Holders are represented by:				
Holder of the position represents:				

Continuation Table 1.12

Powers, powers of attorney:				

Continuation Table 1.12

The following tools belong to the task area	

<table>
<tr><td colspan="3">Machinery, technical equipment:</td><td></td><td></td></tr>
<tr><td colspan="4">The following specialist knowledge is required to be able to perform the task area:</td><td></td></tr>
<tr><td colspan="3"></td><td></td><td></td></tr>
<tr><td></td><td></td><td></td><td></td><td></td></tr>
</table>

2 Self-Management

Form 2.1: Definition of Values

Acclaim	
Accountability	
Achievement	
Beauty & Beauty	
Bravery in battle	
Cash	
Cheerfulness	
Comfort	
Courage	
Desire	
Discipline	
Exhilaration	
Fidelity	
Fortunateness	
Friendship	
Fulfilment of one's duty	

Continuation Form 2.1

Glory	
Harmony	
Health	
Influence	
Joke	
Joy	
Justice	
Love	
Modesty	
Order	
Power	
Power	
Property	
Rigor	
Safety and security	
Solicitude	
Steadfastness of faith	

Continuation Form 2.1

Tact	
Trustworthiness	
Truthfulness	
Wisdom	

Form 2.2: Attitude to Handle Time

The stocktaking of attitudes and the handling of time also includes the feeling of having much or little time.

	Grades (1 best, 5 worst)				
	1	2	3	4	5
My schedule's always full to the brim.					
I would like to have much more free time to take advantage of all my private possibilities.					
My working hours per week are considerably more than forty hours.					
After work I am so exhausted that I cannot develop any more activities.					
I often take work home with me.					
I hardly have time for my private life.					

Continuation Form 2.2

Also in my spare time my thoughts often revolve around professional tasks and problems.					
I often have the feeling that I haven't achieved anything, even though I worked hard all day.					
I want to be reachable all the time, otherwise I have the feeling of missing opportunities.					
I hardly get to things I should be doing.					
There is no time for creativity and self-realization.					

Form 2.3: Self-Assessment of Personal Qualifications

First edit the form alone (self-image) and then let a person close to you (external image) judge you. You will certainly learn a lot from the differences.

If you don't want to address strangers, you can also gradually check the individual points in dealing with others. The reactions of your counterpart (facial expressions, gestures, body language) will show you how to "get along" over time.

Criteria	**Evaluation** (1 best, 5 worst)				
	1	2	3	4	5
Personal Responsibility					
energetic					
dreamily					
goal-oriented					
effective					
Performance motivation					
willing to take risks					
unruffled					
vibrant					
resilient					

Continuation Form 2.3

Self-Confidence

contented					
emotive					
authoritarian					
active					
Motivation					
engaged					
doubting					
obstinate					
dutiful					
Sociability					
convivial					
emotionally					
rational					
extroverted					
Appearance					
dominant					

Continuation Form 2.3

helpful					
cautious					
self-critical					

Sensitivity					
sensitive					
tolerant					
energetic					
ruminative					
Enthusiasm					
sincere					
unblemished					
proactive					
disciplined					
Status Thinking					
material					
popular					
unboastfully					

Continuation Form 2.3

aware of validity					
Systematic Working					
analytically					
spontaneous					
creatively					

organised					
Initiative					
stubborn					
self-reliant					
reactive					
matched					
Tolerance for Frustration					
excitable					
sovereign					
strong-willed					
jocund					
Critical Faculties					

Continuation Form 2.3

sensitive					
diplomatically					
temporising					
humorous					
Emotionality					
optimistic					
circumspect					

reasonable					
motivating					
Self- Assurance					
indemnifying					
demanding					
counterpoising					
familiar					
loyal					
flexible					
critical					
progressive					

3 Team Management

Form 3.1: Evaluation of the Working Atmosphere in the Team

	Theme	does not apply	don´t know	does apply
1	The climate is unbureaucratic, comfortable and relaxed.			
2	The roles, responsibilities and powers have been clarified.			
3	We take the time to share our goals and expectations with each other.			
4	Clear agreements are made and accepted.			
5	The members are willing to cooperate in the realization of the project goals.			
6	We abide by the decisions made and implement them.			
7	We can speak openly about uncertainties and fears.			
8	Differences of opinion are completely clarified and individual points of view are taken into account.			
9	We invest time in controlling our problem solving strategy.			

Continuation Form 3.1

10	Resolutions are made on the basis of			

	consensus.			
11	We talk about the quality of our cooperation.			

Form 3.2: Team Roles According to BELBIN Self-Assessment

What is your best role in the team?

With the following BELBIN test you can answer these questions.

- Read statement with possible answers carefully first.
- Start with "1" and estimate the value of the sentence in relation to your attitudes and behaviour.
- Please distribute 10 points among all statements that apply to you. The more accurate the statement, the more points you give. If you only find yourself in one sentence, it will receive 10 points, otherwise divide the points as you see yourself. The more you divide your points between different questions, the more unclear the result becomes, so have the courage to decide!

Finally, only 10 points should be distributed for each question.

1. I love my work because ...

2. Characteristic for my way of working in a group is ...

A	I find it interesting to get to know my colleagues better in order to get on better with them.
B	I usually put forward arguments to bring down inappropriate pro-

	posals.
C	I think that I can forge captivating action plans which then lead to action.
D	I have the gift to bring unexpected alternatives to the table instead of presenting "old wine in new bottles".
E	When it comes to solving tasks problems in a team, I'm more of a perfectionist.
F	I like to have useful contacts outside the current working group.
G	When it comes to solving a task, I devote myself to it with love for detail.
H	If something has to be tackled, I know right away what has to be done, even if I am interested in everyone's opinions.
I	I am not afraid to discuss against the opinions of others or to defend the position of a minority.

3. When I work with others in a project ...

A	I succeed in guiding people without pushing them in any direction.
B	My vigilance protects us from inaccuracy and from forgetting things.
C	I urge concrete action to ensure that no time is wasted in meetings and to separate the important from the unimportant.
D	You can be sure that I will bring in new and basic ideas.
E	I'm always ready to contribute a good idea if it helps everyone.
F	I am always on the lookout for new ideas and developments.
G	I am convinced that with my broad range of knowledge and my skills I can bring others a good deal closer to the solution.
H	I am convinced that my ability to judge makes an important contribution to decision making.
I	Others can be sure that I handle all the important questions.

4. What I can contribute to a team ...

A	I think that I am able to discover and apply new alternatives.
B	My ability lies in my expertise, which I like to use to advance the task.
C	I can get along with everyone.
D	Throwing in ideas is one of my natural gifts.
E	My ability is to challenge and encourage people when I feel they can make a significant contribution.
F	My efficient work allows me to finish things (e.g. projects) cleanly.
G	I am prepared to be unpopular for a while when it leads to results that are valuable.
H	Normally I realize what is possible and realistic.
I	I can provide reasons for alternative approaches without losing sight of my ultimate goal.

5. When working in a team my weaknesses could be that ...

A	I only feel comfortable when meetings are well structured, under control and well conducted.
B	When people have good ideas on a subject, I agree too quickly without discussing it thoroughly.
C	I run the risk of talking too much when I have new ideas.
D	Normally I don't have a very strong opinion. As a consequence, I quickly agree with others.
E	In situations where something has to happen, I am sometimes perceived as pushy and authoritarian.
F	I find it difficult to take command quickly - perhaps because I am very sensitive to the moods in the group.
G	I have a tendency to "take off" with my ideas and lose contact with what is actually going on.
H	My colleagues think that I unnecessarily waste time with possible failures.
I	I feel comfortable when I can deal in detail with my own task.

6. When suddenly a difficult task is assigned to me, which is to be mastered under time pressure and with new people unknown to me ...

A	I prefer to hold back to sound out the situation before coming up with proposals.
B	I immediately look for ways of working with the people I like the most.
C	I'm immediately looking for a way to tackle the task by determining which people are best to contribute which contribution.
D	My feeling tells me what is more or less urgent and how the schedule should look like.
E	I stay calm and collect my skills to think objectively.
F	I always work towards the goal, even if sometimes only in small steps.
G	I will take the lead if I find that the group is not making progress.
H	I immediately initiate discussions to provoke new perspectives and start the solution process.
I	I immediately integrate into the group and wait for a task to be assigned to me that I can devote myself to.

7. While working in a group, I struggle with the following problems ...

A	I often show impatience towards those who hinder progress.
B	Other people criticize me for being too analytical and not trusting my intuition.
C	My over-concern that things might not be done carefully enough brings with it the risk that project progress will be hampered.
D	It is difficult for me to keep an eye on the overall goal, I prefer to devote myself to fulfilling the partial goals.
E	I quickly get bored - and then I need some impetus to get active again.
F	It is difficult for me to start when the goals are not yet clearly defined.
G	I am hardly able to bring order into the abundance of my thoughts and ideas.
H	I often tend to ask other people to do things I don't like to do myself.
I	I always hesitate to express my thoughts when I think that this will create a stronger opposition.

Evaluation

Question	Points Distribution								
1	G	E	I	A	B	F	D	C	H
2	I	B	E	C	D	G	A	H	F
3	A	C	D	H	I	F	E	G	B
4	E	G	D	I	H	A	C	B	F
5	B	E	G	D	A	C	F	I	H
6	C	G	A	E	F	H	B	I	D
7	H	A	G	B	F	E	I	D	C
Sum									

	A	**B**	**C**	**D**	**E**	**F**	**G**	**H**	**I**
Role	Chairman Coordinator	Implementor	Planter	Monitor Elevator	Shaper	Ressource Investigator	Teamworker	Specialist	Completer Finisher

INDEX OF TERMS

LIST OF TRADEMARKS

Microsoft (MS) Project 2010® ist ein Produkt der Microsoft® Coorp., U. S. A.

Microsoft Excel, Corel DrawTM 7, Oracle, Visual Basic® und ähnliche sind entweder eingetragene Marken oder Marken der Microsoft Corp. und/oder anderer Unternehmen in den Vereinigten Staaten und/oder in anderen Ländern,

MindManager Version 2012, Mindjet LLC.

Trello https://**trello**.com/

APPENDIX: LIST OF LINKS

Free Strategy Session http://bit.ly/2FBysxb

Contact https://www.kisp.de/kontakt

Blog „Selbstführung & Produktivität" https://www.kisp.de/blog

Google-Formular https://forms.gle/NDxWfb2tCTZJGP61A

Bonus material https://www.kisp.de/5cgi

ABOUT THE AUTHOR

Prof. Dr. Annette Kunow taught mechatronics and mechanical engineering at the Bochum University of Applied Sciences for 32 years after several years in industry.

Prof. Dr. Annette Kunow supports people to help themselves and structure them better. In addition companies that want more efficiency in their project management to bring me the money. Last but not least, she stands for startups in the area of Engineering as a Science Angel.

Her passion is to consistently advance things.

Annette Kunow is the author of several books.

Technische Mechanik Statik

Die Technische Mechanik ist eine Kernkompetenz eines jeden Ingenieurs. Ohne diese Kenntnisse können die physikalischen Eigenschaften von Systemen nicht erfasst werden.

Was Sie in diesem Buch lernen werden

1. Mathematische Grundlagen
2. Arbeitsbegriff der Statik
3. Gleichgewicht
4. Schnitt- und Reaktionskräfte
5. Haftung und Reibung
6. Raumstatik

Technische Mechanik Statik Übungen

Die Technische Mechanik ist eine Kernkompetenz eines jeden Ingenieurs. Ohne diese Kenntnisse können die physikalischen Eigenschaften von Systemen nicht erfasst werden.

Vollständig und mit möglichen Lösungsvarianten gelöste Übungsaufgaben

Was Sie in diesem Buch lernen werden

7. Mathematische Grundlagen
8. Arbeitsbegriff der Statik
9. Gleichgewicht
10. Schnitt- und Reaktionskräfte
11. Haftung und Reibung
12. Raumstatik

Technische Mechanik Elastostatik

Die Technische Mechanik ist eine Kernkompetenz eines jeden Ingenieurs. Ohne diese Kenntnisse können die physikalischen Eigenschaften von Systemen nicht erfasst werden.

Was Sie in diesem Buch lernen werden

1. Deformationen
2. Elastizitätsgesetz
3. Spannungen
4. Spannungszustände
5. Statische Bestimmtheit
6. Arbeitsbegriff der Elastostatik

Technische Mechanik Elastostatik Übungen

Die Technische Mechanik ist eine Kernkompetenz eines jeden Ingenieurs. Ohne diese Kenntnisse können die physikalischen Eigenschaften von Systemen nicht erfasst werden.

Vollständig und mit möglichen Lösungsvarianten gelöste Übungsaufgaben

Was Sie in diesem Buch lernen werden

7. Deformationen
8. Elastizitätsgesetz
9. Spannungen
10. Spannungszustände
11. Statische Bestimmtheit
12. Arbeitsbegriff der Elastostatik

Technische Mechanik Dynamik

Die Technische Mechanik ist eine Kernkompetenz eines jeden Ingenieurs. Ohne diese Kenntnisse können die physikalischen Eigenschaften von Systemen nicht erfasst werden.

Was Sie in diesem Buch lernen werden

- Kinematik
- Kinetik des Massenpunktes
- Kinetik des Massenpunktsystems
- Kinetik des Starrkörpers
- Ebene Bewegung
- Schwingungen

Technische Mechanik Dynamik Übungen

Die Technische Mechanik ist eine Kernkompetenz eines jeden Ingenieurs. Ohne diese Kenntnisse können die physikalischen Eigenschaften von Systemen nicht erfasst werden.

Vollständig und mit möglichen Lösungsvarianten gelöste Übungsaufgaben

Was Sie in diesem Buch lernen werden

- Kinematik
- Kinetik des Massenpunktes
- Kinetik des Massenpunktsystems
- Kinetik des Starrkörpers
- Ebene Bewegung
- Schwingungen

Projektmanagement und Business Coaching

Grundlagen des agilen Projektmanagements mit Methoden des Systemischen Coachings

Projektkompetenz ist heute die Kernkompetenz für jeden Berufstätigen. Ohne die Strukturierung durch das Projektmanagement sind Abläufe in Unternehmen nicht mehr zu bewältigen.

Was Sie in diesem Buch lernen werden

- Strukturierte Pläne
- Optimale Nutzung der Ressourcen
- Klar bewertbare Projektziele
- Angepasste Informationssysteme
- Führung des Teams
- Strategische Projektziele

Project Management and Business Coaching

Basics of Agile Project Management With Methods of Systemic Coaching

Project competence is today the core competence for every professional. Without structuring through project management, processes in companies can no longer be mastered.

What You Will Learn in this Book

- Structured plans
- Optimal use of resources
- Clearly assessable project objectives
- Adapted information systems
- Leadership of the team
- Strategic project goals

Finite-Elemente Methode / Computer Aided Engineering (CAE)

Theoretische Grundlagen und Lösungen

CAE ist heute in den Konstruktions- und Entwicklungsbereichen der Industrie nicht mehr wegzudenken. Die heute übliche automatische Vernetzung kann ohne das Grundlagenwissen zu gravierenden Fehlern führen.

Was Sie in diesem Buch lernen werden

- Grundbegriffe und Gesamtsteifigkeit
- Flächen- und Volumenelemente
- Vernetzungsregeln
- Versuche
- Dynamische Berechnungen
- Nichtlinearität

Numerische Dynamik

Grundlagen-Modellbildung-Anwendungen

Die Numerische Dynamik ist ein bedeutender Bestandteil im Engineering. Sie vermittelt die physikalischen Zusammenhänge, um Konstruktionen unter bewegten Belastungen zu dimensionieren.

Was Sie in diesem Buch lernen werden

- Grundlagen der Dynamik/ Kinetik
- Prinzip der dynamischen Berechnung
    - Einmassenschwinger
    - System mit zwei Freiheitsgraden
    - Mehrmassensystem
- Berechnung für das Kontinuum
- Ausführliche Beispiele und Übungen, incl. Eingaben in die Programme (EXCEL, MATLAB)

Numerische Dynamik Übungen

Grundlagen-Modellbildung-Anwendungen

Die Numerische Dynamik ist ein bedeutender Bestandteil im Engineering. Sie vermittelt die physikalischen Zusammenhänge, um Konstruktionen unter bewegten Belastungen zu dimensionieren.

Was Sie in diesem Buch lernen werden

- Grundlagen der Dynamik/Kinetik
- Prinzip der dynamischen Berechnung
    - Einmassenschwinger
    - System mit zwei Freiheitsgraden
    - Mehrmassensystem
- Berechnung für das Kontinuum
- Ausführliche Beispiele und Übungen, incl. Eingaben in die Programme (EXCEL, MATLAB)

24 raketenstarke Produktivitäts-Stategien

So bringen Sie Ihren ganz persönlichen Business-Alltag zum Abheben

In diesem Booklet finden Sie eine Zusammenfassung der wichtigsten Produktivitäts-Strategien.

Was Sie in diesem Buch lernen werden

- Zusammenfassung der wichtigsten Produktivitäts-Strategien

24 Rocket Power Productivity Strategies

How to make your very personal business day stand out

In this booklet you will find a summary of the most important productivity strategies.

What You Will Learn in this Book

- Summary of the most important productivity strategies

7 auftriebsstarke Listen für Ihr Projekt

So bringen Sie Ihr nächstes Projekt in Sekundenschnelle nach oben

Ich zeige Ihnen hier meine 7 wichtigsten Checklisten zur Planung und erfolgreichen Durchführung eines Projektes.

Und Sie bekommen eine umfangreiche Beschreibung zu den einzelnen Schritten, um es gut umsetzen zu können.

Was Sie in diesem Buch lernen werden

- Checklisten zur Planung und erfolgreichen Durchführung eines Projektes mit Beschreibunmhg

7 Buoyant Lists For Your Project

How to bring your next project upwards in a matter of seconds

Here I show you my 7 most important checklists for planning and successful implementation of a project.

And you will receive a comprehensive description of the individual steps in order to be able to implement it well.

What You Will Learn in this Book

- Checklists for planning and successful implementation of a project

www.ingramcontent.com/pod-product-compliance
Lightning Source LLC
Chambersburg PA
CBHW041323210926
51847CB00003B/36